Universitext

[illegible]

Carles Cas[illegible]
[illegible] de Barcelona

Angus Ma[illegible]
[illegible] of London

Kenneth R[illegible]
[illegible]

[illegible]
[illegible]

[illegible]
[illegible]

[illegible]
[illegible]

[illegible] is a series of textbooks that presents material from a wide variety of mathematical disciplines at master's level and beyond. The books, often well class-tested by their author, may have an informal, personal even experimental approach to their subject matter. Some of the most successful and established books in the series have evolved through several editions, always following the evolution of teaching curricula, to very polished texts.

[illegible] trickle down into graduate-level teaching, first textbooks written for new, cutting-edge courses may make their way into Universitext.

For further volumes:
http://www.springer.com/series/223

Universitext

Series Editors:

Sheldon Axler
San Francisco State University

Vincenzo Capasso
Università degli Studi di Milano

Carles Casacuberta
Universitat de Barcelona

Angus MacIntyre
Queen Mary University of London

Kenneth Ribet
University of California, Berkeley

Claude Sabbah
CNRS, École Polytechnique, Paris

Endre Süli
University of Oxford

Wojbor A. Woyczynski
Case Western Reserve University, Cleveland, OH

Universitext is a series of textbooks that presents material from a wide variety of mathematical disciplines at master's level and beyond. The books, often well class-tested by their author, may have an informal, personal even experimental approach to their subject matter. Some of the most successful and established books in the series have evolved through several editions, always following the evolution of teaching curricula, to very polished texts.

Thus as research topics trickle down into graduate-level teaching, first textbooks written for new, cutting-edge courses may make their way into *Universitext*.

For further volumes:
http://www.springer.com/series/223

Ross G. Pinsky

Problems from the Discrete to the Continuous

Probability, Number Theory, Graph Theory, and Combinatorics

Ross G. Pinsky
Department of Mathematics
Technion-Israel Institute of Technology
Haifa, Israel

ISSN 0172-5939 ISSN 2191-6675 (electronic)
ISBN 978-3-319-07964-6 ISBN 978-3-319-07965-3 (eBook)
DOI 10.1007/978-3-319-07965-3
Springer Cham Heidelberg New York Dordrecht London

Library of Congress Control Number: 2014942157

Mathematics Subject Classification (2010): 05A, 05C, 05D, 11N, 60

© Springer International Publishing Switzerland 2014
This work is subject to copyright. All rights are reserved by the Publisher, whether the whole or part of the material is concerned, specifically the rights of translation, reprinting, reuse of illustrations, recitation, broadcasting, reproduction on microfilms or in any other physical way, and transmission or information storage and retrieval, electronic adaptation, computer software, or by similar or dissimilar methodology now known or hereafter developed. Exempted from this legal reservation are brief excerpts in connection with reviews or scholarly analysis or material supplied specifically for the purpose of being entered and executed on a computer system, for exclusive use by the purchaser of the work. Duplication of this publication or parts thereof is permitted only under the provisions of the Copyright Law of the Publisher's location, in its current version, and permission for use must always be obtained from Springer. Permissions for use may be obtained through RightsLink at the Copyright Clearance Center. Violations are liable to prosecution under the respective Copyright Law.
The use of general descriptive names, registered names, trademarks, service marks, etc. in this publication does not imply, even in the absence of a specific statement, that such names are exempt from the relevant protective laws and regulations and therefore free for general use.
While the advice and information in this book are believed to be true and accurate at the date of publication, neither the authors nor the editors nor the publisher can accept any legal responsibility for any errors or omissions that may be made. The publisher makes no warranty, express or implied, with respect to the material contained herein.

Printed on acid-free paper

Springer is part of Springer Science+Business Media (www.springer.com)

In most sciences one generation tears down what another has built and what one has established another undoes. In Mathematics alone each generation builds a new story to the old structure.

—Hermann Hankel

A peculiar beauty reigns in the realm of mathematics, a beauty which resembles not so much the beauty of art as the beauty of nature and which affects the reflective mind, which has acquired an appreciation of it, very much like the latter.

—Ernst Kummer

To Jeanette

and to

E. A. P.
Y. U. P.
L. A. T-P.
M. D. P.

Preface

It is often averred that two contrasting cultures coexist in mathematics—the theory-building culture and the problem-solving culture. The present volume was certainly spawned by the latter. This book takes an array of specific problems and solves them, with the needed tools developed along the way in the context of the particular problems.

The book is an unusual hybrid. It treats a mélange of topics from combinatorial probability theory, multiplicative number theory, random graph theory, and combinatorics. Objectively, what the problems in this book have in common is that they involve the asymptotic analysis of a discrete construct, as some natural parameter of the system tends to infinity. Subjectively, what these problems have in common is that both their statements and their solutions resonate aesthetically with me.

The *results* in this book lend themselves to the title "Problems from the Finite to the Infinite"; however, with regard to the *methods of proof*, the chosen appellation is the more apt. In particular, generating functions in their various guises are a fundamental bridge "from the discrete to the continuous," as the book's title would have it; such functions work their magic often in these pages. Besides bridging discrete mathematics and mathematical analysis, the book makes a modest attempt at bridging disciplines—probability, number theory, graph theory, and combinatorics.

In addition to the considerations mentioned above, the problems were selected with an eye toward accessibility to a wide audience, including advanced undergraduate students. The technical prerequisites for the book are a good grounding in basic undergraduate analysis, a touch of familiarity with combinatorics, and a little basic probability theory. One appendix provides the necessary probabilistic background, and another appendix provides a warm-up for dealing with generating functions. That said, a moderate dose of the elusive quality known as mathematical maturity will certainly be helpful throughout the text and will be necessary on occasion.

The primary intent of the book is to introduce a number of beautiful problems in a variety of subjects quickly, pithily, and completely rigorously to graduate students and advanced undergraduates. The book could be used for a seminar/capstone course in which students present the lectures. It is hoped that the book might also be of interest to mathematicians whose fields of expertise are away from the subjects treated herein. In light of the primary intended audience, the level of detail in proofs is a bit greater than what one sometimes finds in graduate mathematics texts.

I conclude with some brief comments on the novelty or lack thereof in the various chapters. A bit more information in this vein may be found in the chapter notes at the end of each chapter. Chapter 1 follows a standard approach to the problem it solves. The same is true for Chap. 2 except for the probabilistic proof of Theorem 2.1, which I haven't seen in the literature. The packing problem result in Chap. 3 seems to be new, and the proof almost certainly is. My approach to the arcsine laws in Chap. 4 is somewhat different than the standard one; it exploits generating functions to the hilt and is almost completely combinatorial. The traditional method of proof is considerably more probabilistic. The proofs of the results in Chap. 5 on the distribution of cycles in random permutations are almost exclusively combinatorial, through the method of generating functions. In particular, the proof of Theorem 5.2 makes quite sophisticated use of this technique. In the setting of weighted permutations, it seems that the method of proof offered here cannot be found elsewhere. The number theoretic topics in Chaps. 6–8 are developed in a standard fashion, although the route has been streamlined a bit to provide a rapid approach to the primary goal, namely, the proof of the Hardy–Ramanujan theorem. In Chap. 9, the proof concerning the number of cliques in a random graph is more or less standard. The result on tampering detection constitutes material with a new twist and the methods are rather probabilistic; a little additional probabilistic background and sophistication on the part of the reader would be useful here. The results from Ramsey theory are presented in a standard way. Chapter 10, which deals with the phase transition concerning the giant component in a sparse random graph, is the most demanding technically. The reader with a modicum of probabilistic sophistication will be at quite an advantage here. It appears to me that a complete proof of the main results in this chapter, with all the details, is not to be found in the literature.

Acknowledgements It is a pleasure to thank my editor, Donna Chernyk, for her professionalism and superb diligence.

Haifa, Israel
April 2014

Ross G. Pinsky

Contents

A Note on Notation

$\mathbb{Z}$ denotes the set of integers

$\mathbb{Z}^+$ denotes the set of nonnegative integers

$\mathbb{N}$ denotes the set of natural numbers: $\{1, 2, \cdots\}$

$\mathbb{R}$ denotes the set of real numbers

$f(x) = O(g(x))$ as $x \to a$ means that $\limsup_{x \to a} |\frac{f(x)}{g(x)}| < \infty$; in particular, $f(x) = O(1)$ as $x \to a$ means that $f(x)$ remains bounded as $x \to a$

$f(x) = o(g(x))$ as $x \to a$ means that $\lim_{x \to a} \frac{f(x)}{g(x)} = 0$; in particular, $f(x) = o(1)$ as $x \to a$ means $\lim_{x \to a} f(x) = 0$

$f \sim g$ as $x \to a$ means that $\lim_{x \to a} \frac{f(x)}{g(x)} = 1$

$\gcd(x_1, \ldots, x_m)$ denotes the greatest common divisor of the positive integers $x_1, \ldots, x_m$

The symbol $[\cdot]$ is used in two contexts:

1. $[n] = \{1, 2, \ldots, n\}$, for $n \in \mathbb{N}$
2. $[x]$ is the greatest integer function; that is, $[x] = n$, if $n \in \mathbb{Z}$ and $n \le x < n + 1$

$\mathrm{Bin}(n, p)$ is the binomial distribution with parameters n and p

$\mathrm{Pois}(\lambda)$ is the Poisson distribution with parameter λ

$\mathrm{Ber}(p)$ is the Bernoulli distribution with parameter p

$X \sim \mathrm{Bin}(n, p)$ means the random variable X is distributed according to the distribution $\mathrm{Bin}(n, p)$

A Note on Notation

Chapter 1
Partitions with Restricted Summands or "the Money Changing Problem"

Imagine a country with coins of denominations 5 cents, 13 cents, and 27 cents. How many ways can you make change for \$51,419.48? That is, how many solutions (b_1, b_2, b_3) are there to the equation $5b_1 + 13b_2 + 27b_3 = 5{,}141{,}948$, with the restriction that b_1, b_2, b_3 be nonnegative integers? This is a specific case of the following general problem. Fix m distinct, positive integers $\{a_j\}_{j=1}^m$. Count the number of solutions $(b_1, \ldots, b_m)$ with integral entries to the equation

$$b_1a_1 + b_2a_2 + \cdots + b_ma_m = n, \; b_j \geq 0, \; j = 1, \ldots, m. \tag{1.1}$$

A *partition* of n is a sequence of integers $(x_1, \ldots, x_k)$, where k is a positive integer, such that

$$\sum_{i=1}^{k} x_i = n \text{ and } x_1 \geq x_2 \geq \cdots \geq x_k \geq 1.$$

Let P_n denote the number of different partitions of n. The problem of obtaining an asymptotic formula for P_n is celebrated and very difficult. It was solved in 1918 by G.H. Hardy and S. Ramanujan, who proved that

$$P_n \sim \frac{1}{4n\sqrt{3}} e^{\pi\sqrt{\frac{2n}{3}}}, \text{ as } n \to \infty.$$

Now consider partitions of n where we restrict the values of the summands x_i above to the set $\{a_j\}_{j=1}^m$. Denote the number of such *restricted partitions* by $P_n(\{a_j\}_{j=1}^m)$. A moment's thought reveals that the number of solutions to (1.1) is $P_n(\{a_j\}_{j=1}^m)$.

Does there exist a solution to (1.1) for every sufficiently large integer n? And if so, can one evaluate asymptotically the number of such solutions for large n? Without posing any restrictions on $\{a_j\}_{j=1}^m$, the answer to the first question is negative. For example, if $m = 3$ and $a_1 = 5, a_2 = 10, a_3 = 30$, then clearly there is no solution to (1.1) if $n \nmid 5$. Indeed, it is clear that a necessary condition

R.G. Pinsky, *Problems from the Discrete to the Continuous*, Universitext,
DOI 10.1007/978-3-319-07965-3_1, © Springer International Publishing Switzerland 2014

for the existence of a solution for all large n is that $\{a_j\}_{j=1}^m$ are relatively prime: $\gcd(a_1,\ldots,a_m) = 1$. This is the time to recall a well-known result concerning solutions $(b_1,\ldots,b_m)$ with (not necessarily nonnegative) integral entries to the equation $b_1a_1+b_2a_2+\cdots+b_ma_m = n$. A fundamental theorem in algebra/number theory states that there exists an integral solution to this equation for all $n \in \mathbb{Z}$ if and only if $\gcd(a_1,\ldots,a_m) = 1$. This result has an elegant group theoretical proof. We will prove that for all large n, (1.1) has a solution $(b_1,\ldots,b_m)$ with integral entries if and only if $\gcd(a_1,\ldots,a_m) = 1$, and we will give a precise asymptotic estimate for the number of such solutions for large n.

Theorem 1.1. *Let $m \geq 2$ and let $\{a_j\}_{j=1}^m$ be distinct, positive integers. Assume that the greatest common divisor of $\{a_j\}_{j=1}^m$ is 1: $\gcd(a_1,\ldots,a_m) = 1$. Then for all sufficiently large n, there exists at least one integral solution to* (1.1). *Furthermore, the number $P_n(\{a_j\}_{j=1}^m)$ of such solutions satisfies*

$$P_n(\{a_j\}_{j=1}^m) \sim \frac{n^{m-1}}{(m-1)!\prod_{j=1}^m a_j}, \text{ as } n \to \infty. \tag{1.2}$$

Remark. In particular, we see (not surprisingly) that for fixed m and sufficiently large n, the smaller the $\{a_j\}_{j=1}^m$ are, the more solutions there are. We also see that given m_1 and $\{a_j^{(1)}\}_{j=1}^{m_1}$, and given m_2 and $\{a_j^{(2)}\}_{j=1}^{m_2}$, with $m_2 > m_1$, then for sufficiently large n there will be more solutions for the latter set of parameters.

Proof. We will prove the asymptotic estimate in (1.2), from which the first statement of the theorem will also follow. Let h_n denote the number of solutions to (1.1). (For the proof, the notation h_n will be a lot more convenient than $P_n(\{a_j\}_{j=1}^m)$.) Thus, we need to show that (1.2) holds with h_n in place of $P_n(\{a_j\}_{j=1}^m)$. We define the generating function of $\{h_n\}_{n=1}^\infty$:

$$H(x) = \sum_{n=1}^\infty h_n x^n. \tag{1.3}$$

A simple, rough estimate shows that $h_n \leq \frac{n^m}{\prod_{j=1}^m a_j}$, from which it follows that the power series on the right hand side of (1.3) converges for $|x| < 1$. See Exercise 1.1. It turns out that we can exhibit H explicitly. We demonstrate this for the case $m = 2$, from which the general case will become clear.

For $k = 1, 2$, we have

$$\frac{1}{1-x^{a_k}} = 1 + x^{a_k} + x^{2a_k} + x^{3a_k} + \cdots,$$

and the series converges absolutely for $|x| < 1$. Thus,

$$\frac{1}{(1-x^{a_1})(1-x^{a_2})}=(1+x^{a_1}+x^{2a_1}+x^{3a_1}+\cdots)(1+x^{a_2}+x^{2a_2}+x^{3a_2}+\cdots)=$$
$$(1+x^{a_1}+x^{2a_1}+x^{3a_1}+\cdots)+(x^{a_2}+x^{a_1+a_2}+x^{2a_1+a_2}+x^{3a_1+a_2}+\cdots)+$$
$$(x^{2a_2}+x^{a_1+2a_2}+x^{2a_1+2a_2}+x^{3a_1+2a_2}+\cdots)+\cdots \tag{1.4}$$

A little thought now reveals that on the right hand side of (1.4), *the number of times the term x^n appears is the number of integral solutions (b_1, b_2) to (1.1) with $m = 2$*; that is, h_n is the coefficient of x^n on the right hand side of (1.4). So $H(x) = \frac{1}{(1-x^{a_1})(1-x^{a_2})}$. Clearly, the same argument works for all m; thus we conclude that

$$H(x) = \frac{1}{(1-x^{a_1})(1-x^{a_2})\cdots(1-x^{a_m})}, \ |x| < 1. \tag{1.5}$$

We now begin an analysis of H, as given in its closed form in (1.5), which will lead us to the asymptotic behavior as $n \to \infty$ of the coefficients h_n in its power series representation in (1.3). Consider the polynomial

$$p(x) = (1-x^{a_1})(1-x^{a_2})\cdots(1-x^{a_m}).$$

For each k, the roots of $1-x^{a_k}$ are the a_kth roots of unity: $\{e^{\frac{2\pi ij}{a_k}}\}_{j=0}^{a_k-1}$. Clearly 1 is a root of $p(x)$ of multiplicity m. *Because of the assumption that $gcd(a_1,\ldots,a_m) = 1$, it follows that every other root of $p(x)$ is of multiplicity less than m*—that is, there is no complex number r that can be written in the form $r = e^{\frac{2\pi ij_k}{a_k}}$, simultaneously for $k = 1,\ldots,m$, where $1 \le j_k < a_k$. Indeed, if r can be written in the above form for all k, then it follows that $\frac{j_k}{a_k}$ is independent of k. In particular, $a_k = \frac{j_k a_1}{j_1}$, for $k = 2,\ldots,m$. Since $1 \le j_1 < a_1$, it follows that there is at least one prime factor of a_1 which is a factor of all of the a_k, $k = 2,\ldots,m$, and this contradicts the assumption that $\gcd(a_1,\ldots,a_m) = 1$.

Denote the distinct roots of $p(x)$ by $1, r_2,\ldots,r_l$, and note from above that $|r_j| = 1$, for all j. Let m_k denote the multiplicity of the root r_k, for $k = 2,\ldots,l$. Also, note that $p(0) = 1$. Then we can write

$$(1-x^{a_1})(1-x^{a_2})\cdots(1-x^{a_m}) = (1-x)^m(1-\frac{x}{r_2})^{m_2}\cdots(1-\frac{x}{r_l})^{m_l}, \tag{1.6}$$

where $1 \le m_j < m$, for $j = 2,\ldots,l$.

In light of (1.5) and (1.6), we can write the generating function $H(x)$ in the form

$$H(x) = \frac{1}{(1-x)^m(1-\frac{x}{r_2})^{m_2}\cdots(1-\frac{x}{r_l})^{m_l}}. \tag{1.7}$$

By the method of partial fractions, we can rewrite H from (1.7) in the form

$$H(x) = \Big(\frac{A_{11}}{(1-x)^m} + \frac{A_{12}}{(1-x)^{m-1}} + \cdots + \frac{A_{1m}}{(1-x)}\Big) +$$
$$\Big(\frac{A_{21}}{(1-\frac{x}{r_2})^{m_2}} + \cdots + \frac{A_{2m_2}}{(1-\frac{x}{r_2})}\Big) + \cdots + \Big(\frac{A_{l1}}{(1-\frac{x}{r_l})^{m_l}} + \cdots + \frac{A_{lm_l}}{(1-\frac{x}{r_l})}\Big). \quad (1.8)$$

For positive integers k, the function $F(x) = (1-x)^{-k}$ has the power series expansion

$$(1-x)^{-k} = \sum_{n=0}^{\infty} \binom{n+k-1}{k-1} x^n.$$

To prove this, just verify that $\frac{F^{(n)}(0)}{n!} = \binom{n+k-1}{k-1}$. Thus, the first term on the right hand side of (1.8) can be expanded as

$$\frac{A_{11}}{(1-x)^m} = A_{11} \sum_{n=0}^{\infty} \binom{n+m-1}{m-1} x^n. \quad (1.9)$$

The coefficient of x^n on the right hand side above is

$$A_{11} \frac{(n+m-1)(n+m-2)\cdots(n+1)}{(m-1)!} \sim A_{11} \frac{n^{m-1}}{(m-1)!} \text{ as } n \to \infty.$$

Every other term on the right hand side of (1.8) is of the form $\frac{A}{(1-\frac{x}{r})^k}$ where $1 \le k < m$ and $|r| = 1$. By the same argument as above, the coefficient of x^n in the expansion for $\frac{A}{(1-\frac{x}{r})^k}$ is asymptotic to $\frac{An^{k-1}}{r^n(k-1)!}$ as $n \to \infty$ (substitute $\frac{x}{r}$ for x in the appropriate series expansion). Thus, each of these terms is on a smaller order than the coefficient of x^n in (1.9). We thereby conclude that the coefficient of x^n in $H(x)$ is asymptotic to $A_{11}\frac{n^{m-1}}{(m-1)!}$ as $n \to \infty$. By (1.3), this gives

$$h_n \sim A_{11} \frac{n^{m-1}}{(m-1)!}, \text{ as } n \to \infty. \quad (1.10)$$

It remains to evaluate the constant A_{11}. From (1.8), it follows that

$$H(x) \sim \frac{A_{11}}{(1-x)^m} + O\Big(\frac{1}{(1-x)^{m-1}}\Big), \text{ as } x \to 1.$$

Thus,

$$\lim_{x\to 1} (1-x)^m H(x) = A_{11}. \quad (1.11)$$

But on the other hand, from (1.5), we have

$$(1-x)^m H(x) = \frac{(1-x)^m}{(1-x^{a_1})(1-x^{a_2})\cdots(1-x^{a_m})} = \prod_{j=1}^{m} \frac{x-1}{x^{a_j}-1}. \tag{1.12}$$

Since $(x^{a_j})'|_{x=1} = a_j x^{a_j-1}|_{x=1} = a_j$, we conclude from (1.12) that

$$\lim_{x\to 1}(1-x)^m H(x) = \frac{1}{\prod_{j=1}^{m} a_j}. \tag{1.13}$$

From (1.11) and (1.13) we obtain $A_{11} = \frac{1}{\prod_{j=1}^{m} a_j}$, and thus from (1.10) we conclude that $h_n \sim \frac{n^{m-1}}{(m-1)!\prod_{j=1}^{m} a_j}$. □

Exercise 1.1. If $b_1a_1 + b_2a_2 + \cdots + b_ma_m = n$, then of course $b_ja_j \le n$, for all $j \in [m]$. Use this to obtain the following rough upper bound on the number of solutions h_n to (1.1): $h_n \le \frac{n^m}{\prod_{j=1}^{m} a_j}$. Then use this estimate together with the third "fundamental result" in Appendix B to show that the series defining $H(x)$ in (1.3) converges for $|x| < 1$.

Exercise 1.2. Go through the proof of Theorem 1.1 and convince yourself that the result of the theorem holds even if the integers $\{a_j\}_{j=1}^m$ are not distinct. That is, the number of solutions to (1.1) is asymptotic to the expression on the right hand side of (1.2). Note though that the number of such solutions is not equal to $P_n(\{a_j\}_{j=1}^n)$. What is the leading asymptotic term as $n \to \infty$ for the number of ways to make n cents out of quarters and pennies, where one distinguishes the quarters by their mint marks—"p" for Philadelphia, "d" for Denver, and "s" for San Francisco—but where the pennies are not distinguished?

Exercise 1.3. In the case that $d = \gcd(a_1,\ldots,a_m) > 1$, use Theorem 1.1 to formulate and prove a corresponding result.

Exercise 1.4. A *composition* of n is an ordered sequence of positive integers $(x_1,\ldots,x_k)$, where k is a positive integer, such that $\sum_{i=1}^{k} x_i = n$. A favorite method of combinatorists to calculate the size of some combinatorial object is to find a bijection between the object in question and some other object whose size is known. Let C_n denote the number of compositions of n. To calculate C_n, we construct a bijection as follows. Consider a sequence of n dots in a row. Between each pair of adjacent dots, choose to place or choose not to place a vertical line. Consider the set of all possible dot and line combinations. (For example, if $n = 5$, here are two possible such combinations: (1) $\cdot\cdot|\cdot|\cdot\cdot$ (2) $\cdot\cdot\cdot\cdot\cdot$):

(a) Show that there are 2^{n-1} dot and line combinations.
(b) Show that there is a bijection between the set of compositions of n and the set of dot and line combinations.
(c) Conclude from (a) and (b) that $C_n = 2^{n-1}$.

Exercise 1.5. Let $C_n^{\{1,2\}}$ denote the number of compositions of n with summands restricted to the integers 1 and 2, that is, compositions $(x_1, \cdots, x_k)$ of n with the restriction that $x_i \in \{1,2\}$, for all i. The series

$$F(x) := \frac{1}{1-x-x^2} = \sum_{n=0}^{\infty}(x+x^2)^n \tag{1.14}$$

converges absolutely for $|x| < \frac{\sqrt{5}-1}{2}$ since $|x+x^2| \leq |x| + |x|^2 < 1$ if $|x| < \frac{\sqrt{5}-1}{2}$:

(a) Similar to the argument leading from (1.3) to (1.5), argue that $C_n^{\{1,2\}}$ is the coefficient of x^n in the power series expansion of F.
(b) Show that $F(x) = \sum_{n=0}^{\infty} f_n x^n$, where $\{f_n\}_{n=0}^{\infty}$ is the Fibonacci sequence—see (B.2) in Appendix B. (Hint: One has $(x+x^2)F(x) = F(x) - 1$.)
(c) Conclude from (a) and (b) that $C_n^{\{1,2\}}$ is the nth Fibonacci number; thus, from (B.10) in Appendix B,

$$C_n^{\{1,2\}} = \frac{1}{\sqrt{5}}\Big(\big(\frac{1+\sqrt{5}}{2}\big)^n - \big(\frac{1-\sqrt{5}}{2}\big)^n\Big).$$

Chapter Notes

For a leisurely and folksy introduction to the use of generating functions in combinatorics, see Wilf's little book [34]. For a recent encyclopedic treatment, see the book of Flajolet and Sedgewick [20]. The asymptotic formula for P_n, noted at the beginning of the chapter, was proved by Hardy and Ramanujan in [23]. For a modern account, see [4]. The asymptotic estimate in Theorem 1.1 is due to I. Schur. As noted in the text, this asymptotic formula also proves that (1.1) has a solution for all sufficiently large n. However, this latter fact can be proved more easily; see, for example, Brauer [11]. Given $\{a_j\}_{j=1}^m$, what is the exact minimal value of n_0 such that every $n \geq n_0$ can be written in the form (1.1)? When $m = 2$, the answer is $(a_1 - 1)(a_2 - 1)$. A proof can be found in [34]. For $m \geq 3$ the answer is not known.

Chapter 2
The Asymptotic Density of Relatively Prime Pairs and of Square-Free Numbers

Pick a positive integer at random. What is the probability of it being even? As stated, this question is not well posed, because there is no uniform probability measure on the set $\mathbb{N}$ of positive integers. However, what one can do is fix a positive integer n, and choose a number uniformly at random from the finite set $[n] = \{1, \ldots, n\}$. Letting ρ_n denote the probability that the chosen number was even, we have $\lim_{n\to\infty} \rho_n = \frac{1}{2}$, and we say that the *asymptotic density* of even numbers is equal to $\frac{1}{2}$.

In this spirit, we ask: *if one selects two positive integers at random, what is the probability that they are relatively prime?* Fixing n, we choose two positive integers uniformly at random from $[n]$. Of course, there are two natural ways to interpret this. Do we choose a number uniformly at random from $[n]$ and then choose a second number uniformly at random from the remaining $n - 1$ integers, or, alternatively, do we select the second number again from $[n]$, thereby allowing for doubles? The answer is that it doesn't matter, because under the second alternative the probability of getting doubles is only $\frac{1}{n}$, and this doesn't affect the asymptotic probability. Here is the theorem we will prove.

Theorem 2.1. *Choose two integers uniformly at random from $[n]$. As $n \to \infty$, the asymptotic probability that they are relatively prime is $\frac{6}{\pi^2} \approx 0.6079$.*

We will give two very different proofs of Theorem 2.1: one completely number theoretic and one completely probabilistic. The number theoretic proof is elegant even a little magical. However, it does require the preparation of some basic number theoretic tools, and it provides little intuition. The number theoretic proof gives the asymptotic probability as $(\sum_{n=1}^{\infty} \frac{1}{n^2})^{-1}$. The well-known fact that $\sum_{n=1}^{\infty} \frac{1}{n^2} = \frac{\pi^2}{6}$ is proved in Appendix D. The probabilistic proof requires very little preparation; it is enough to know just the most rudimentary notions from discrete probability theory: probability space, event, and independence. A heuristic, non-rigorous version of the probabilistic proof provides a lot of intuition, some of which the reader might find obscured in the rigorous proof. The probabilistic proof gives the asymptotic probability as $\prod_{k=1}^{\infty}(1 - \frac{1}{p_k^2})$, where $\{p_k\}_{k=1}^{\infty}$ is an enumeration of the primes. One

R.G. Pinsky, *Problems from the Discrete to the Continuous*, Universitext,
DOI 10.1007/978-3-319-07965-3_2, © Springer International Publishing Switzerland 2014

then must use the Euler product formula to show that this is equal to $(\sum_{n=1}^{\infty} \frac{1}{n^2})^{-1}$. We will first give the number theoretic proof and then give the heuristic and the rigorous probabilistic proofs.

The number theoretic ideas we develop along the way to our first proof of Theorem 2.1 will bring us close to proving another result, which we now describe. Every positive integer $n \geq 2$ can be factored uniquely as $n = p_1^{k_1} \cdots p_m^{k_m}$, where $m \geq 1$, $\{p_j\}_{j=1}^m$ are distinct primes, and $k_j \in \mathbb{N}$, for $j \in [m]$. If in this factorization, one has $k_j = 1$, for all $j \in [m]$, then we say that n is *square-free*. Thus, an integer $n \geq 2$ is square-free if and only if it is of the form $n = p_1 \cdots p_m$, where $m \geq 1$ and $\{p_j\}_{j=1}^m$ are distinct primes. The integer 1 is also called square-free. There are 61 square-free positive integers that are no greater than 100: 1,2,3,5,6,7,10,11,13,14,15,17,19,21,22,23,26,29,30,31,33,34,35,37,38,39,41,42,43, 46,47,51,53,55,57,58,59,61,62,65,66,67,69,70,71,73,74,77,78,79,82,83,85,86, 87,89,91,93,94,95,97.

Let $C_n = \{k : 1 \leq k \leq n,\ k \text{ is square-free}\}$. If $\lim_{n\to\infty} \frac{|C_n|}{n}$ exists, we call this limit the asymptotic density of square-free numbers. After giving the number theoretic proof of Theorem 2.1, we will prove the following theorem.

Theorem 2.2. *The asymptotic density of square-free integers is* $\frac{6}{\pi^2} \approx 0.6079$.

For the number theoretic proof of Theorem 2.1, the first alternative suggested above in the second paragraph of this chapter will be more convenient. In fact, once we have chosen the two distinct integers, it will be convenient to order them by size; thus, we may consider the set B_n of all possible (and equally likely) outcomes to be

$$B_n = \{(j,k) : 1 \leq j < k \leq n\}.$$

Let $A_n \subset B_n$ denote those pairs which are relatively prime:

$$A_n = \{(j,k) : 1 \leq j < k \leq n,\ \gcd(j,k) = 1\}.$$

Then the probability q_n that the two selected integers are relatively prime is

$$q_n = \frac{|A_n|}{|B_n|} = \frac{2|A_n|}{n(n-1)}. \tag{2.1}$$

We proceed to develop a circle of ideas that will facilitate the calculation of $\lim_{n\to\infty} q_n$ and thus give a proof of Theorem 2.1. A function $a : \mathbb{N} \to \mathbb{R}$ is called an *arithmetic* function. The *Möbius function* μ is the arithmetic function defined by

$$\mu(n) = \begin{cases} 1, \text{ if } n = 1; \\ (-1)^m, \text{ if } n = \prod_{j=1}^m p_j, \text{ where } \{p_j\}_{j=1}^m \text{ are distinct primes;} \\ 0, \text{ otherwise.} \end{cases}$$

Thus, for example, we have $\mu(3) = -1$, $\mu(15) = 1$, and $\mu(12) = 0$.

Given arithmetic functions a and b, we define their *convolution* $a * b$ to be the arithmetic function satisfying

$$(a * b)(n) = \sum_{d|n} a(d)b(\frac{n}{d}),\ n \in \mathbb{N}.$$

Clearly, $a * b = b * a$. The convolution arises naturally in the following context. Define formally

$$f(x) = \sum_{n=1}^{\infty} \frac{a(n)}{n^x} \tag{2.2}$$

and

$$g(x) = \sum_{n=1}^{\infty} \frac{b(n)}{n^x}. \tag{2.3}$$

When we say "formally," what we mean is that we ignore questions of convergence and manipulate these infinite series according to the laws of addition, subtraction, multiplication, and division, which are valid for series with a finite number of terms and for absolutely convergent infinite series. Their formal product is given by

$$f(x)g(x)=(\sum_{d=1}^{\infty} \frac{a(d)}{d^x})(\sum_{k=1}^{\infty} \frac{b(k)}{k^x})= \sum_{d,k=1}^{\infty} \frac{a(d)b(k)}{(dk)^x}=\sum_{n=1}^{\infty} \frac{1}{n^x} \sum_{d,k:dk=n} a(d)b(k)$$

$$= \sum_{n=1}^{\infty} \frac{1}{n^x} \sum_{d|n} a(d)b(\frac{n}{d}) = \sum_{n=1}^{\infty} \frac{(a * b)(n)}{n^x}. \tag{2.4}$$

If the series on the right hand side of (2.2) and (2.3) are in fact absolutely convergent, then the series on the right hand side of (2.4) is also absolutely convergent. In such case, the equality $\left(\sum_{d=1}^{\infty} \frac{a(d)}{d^x}\right)\left(\sum_{k=1}^{\infty} \frac{b(k)}{k^x}\right) = \sum_{n=1}^{\infty} \frac{(a*b)(n)}{n^x}$ is a rigorous statement in mathematical analysis.

An arithmetic function a is called *multiplicative* if $a(nm) = a(n)a(m)$ whenever $\gcd(n, m) = 1$. It follows that if $a \not\equiv 0$ is multiplicative, then $a(1) = 1$. If $a \not\equiv 0$ is multiplicative, then it is completely determined by its values on the prime powers; indeed, if $n = \prod_{j=1}^{m} p_j^{k_j}$ is the factorization of n into a product of distinct prime powers, then $a(n) = a(\prod_{j=1}^{m} p_j^{k_j}) = \prod_{j=1}^{m} a(p_j^{k_j})$.

It is trivial to verify that μ is multiplicative. For the first proposition below, the following lemma will be useful.

Lemma 2.1. *The arithmetic function* $\sum_{d|n} \mu(d)$ *is multiplicative.*

Proof. Let n and m be positive integers satisfying $\gcd(n, m) = 1$. We have

$$\sum_{d_1|n}\mu(d_1)\sum_{d_2|m}\mu(d_2) = \sum_{d_1|n,d_2|m}\mu(d_1)\mu(d_2) = \sum_{d_1|n,d_2|m}\mu(d_1d_2) = \sum_{d|nm}\mu(d),$$

where the second equality follows from the fact that μ is multiplicative and the fact that if $\gcd(n,m) = 1$, $d_1|n$ and $d_2|m$, then $\gcd(d_1,d_2) = 1$, while the final equality follows from the fact that if $\gcd(n,m) = 1$ and $d|nm$, then d can be written as $d = d_1d_2$ for a unique pair d_1, d_2 satisfying $d_1|n$ and $d_2|m$. (The reader should verify these facts.) □

We introduce three more arithmetic functions that will be used in the sequel:

$$1(n) = 1, \text{ for all } n; \quad i(n) = n, \text{ for all } n; \quad e(n) = \begin{cases} 1, & \text{if } n = 1; \\ 0, & \text{otherwise.} \end{cases}$$

Note that $a * e = a$, for all a, and that $(a * 1)(n) = \sum_{d|n} a(d)$. A key result we need is the *Möbius inversion formula.*

Proposition 2.1. *Let a be an arithmetic function. Define $b = a * 1$. Then $a = b * \mu$.*

Remark. Written out explicitly, the theorem asserts that if

$$b(n) := \sum_{d|n} a(d),$$

then $a(n) = \sum_{d|n} b(d)\mu(\frac{n}{d})$.

Proof. To prove the proposition, it suffices to prove that

$$1 * \mu = e. \tag{2.5}$$

Indeed, using this along with the easily verified associativity of the convolution, we have

$$b * \mu = (a * 1) * \mu = a * (1 * \mu) = a * e = a.$$

We now prove (2.5). We have

$$(1 * \mu)(n) = (\mu * 1)(n) = \sum_{d|n}\mu(d).$$

By Lemma 2.1, the function $\sum_{d|n}\mu(d)$ is multiplicative. Clearly, the function e is multiplicative. Obviously, $e(1) = 1$ and $e(p^k) = 0$, for any prime p and any positive integer k. We have $\sum_{d|1}\mu(d) = \mu(1) = 1$. Thus, since a nonzero, multiplicative, arithmetic function is completely determined by its values on prime powers, to complete the proof that $1 * \mu = e$, it suffices to show that $\sum_{d|p^k} \mu(d) = 0$. We have $\sum_{d|p^k} \mu(d) = \sum_{j=0}^{k} \mu(p^j) = \mu(1) + \mu(p) = 1 - 1 = 0$. □

We introduce one final arithmetic function—the well-known *Euler ϕ-function*:

$$\phi(n) = |\{j : 1 \le j \le n,\ \gcd(j,n) = 1\}|.$$

That is, $\phi(n)$ counts the number of positive integers less than or equal to n which are relatively prime to n. For our calculation of $\lim_{n\to\infty} q_n$, we will use a result that is a corollary of the following proposition.

Proposition 2.2. *$\phi * 1 = i$; that is,*

$$\sum_{d|n} \phi(d) = n.$$

From Proposition 2.2 and the Möbius inversion formula, the following corollary is immediate.

Corollary 2.1. *$\mu * i = \phi$; that is,*

$$\phi(n) = \sum_{d|n} \mu(d)\frac{n}{d}.$$

Remark. For the proofs of Theorems 2.1 and 2.2, we do not need Proposition 2.2, but only Corollary 2.1. In Exercise 2.1, the reader is guided through a direct proof of the corollary. The proof also will reveal why the seemingly strange Möbius function has such nice properties.

Proof of Proposition 2.2. Let $d|n$. It is easy to see that $\phi(d)$ is equal to the number of $k \in [n]$ satisfying $\gcd(k,n) = \frac{n}{d}$. Indeed, $k \in [n]$ satisfies $\gcd(k,n) = \frac{n}{d}$ if and only if $k = j(\frac{n}{d})$, for some $j \in [d]$ satisfying $\gcd(d,j) = 1$. (The reader should verify this.) Also, clearly, every $k \in [n]$ satisfies $\gcd(k,n) = \frac{n}{d}$ for some $d|n$. The proposition follows from these facts. □

Remark. For an alternative proof of Proposition 2.2, exactly in the spirit of Lemma 2.1 and the proof of (2.5), see Exercise 2.2.

We are now in a position to prove Theorem 2.1.

Number Theoretic Proof of Theorem 2.1. For each $k \ge 2$, there are $\phi(k)$ integers j satisfying $1 \le j < k$ and $\gcd(j,k) = 1$. Thus,

$$|A_n| = |\{(j,k) : 1 \le j < k \le n,\ \gcd(j,k) = 1\}| = \sum_{k=2}^{n} \phi(k).$$

Therefore, from (2.1), we have

$$q_n = \frac{2\sum_{k=2}^{n}\phi(k)}{n(n-1)}.$$

To calculate

$$\lim_{n\to\infty} q_n = \lim_{n\to\infty} \frac{2\sum_{k=2}^{n}\phi(k)}{n(n-1)}, \tag{2.6}$$

we analyze the behavior of the sum $\sum_{k=1}^{n}\phi(k)$ for large n.

Remark. The function ϕ can be written explicitly as

$$\phi(n) = n\prod_{p|n}(1-\frac{1}{p}),\ n \geq 2, \tag{2.7}$$

where $\prod_{p|n}$ indicates that the product is over all primes that divide n; see Exercise 2.3. However, this formula is of no help whatsoever for analyzing the above sum.

We will use Corollary 2.1 to analyze $\sum_{k=1}^{n}\phi(k)$. From Corollary 2.1, we have

$$\sum_{k=1}^{n}\phi(k) = \sum_{k=1}^{n}(\mu * i)(k) = \sum_{k=1}^{n}\sum_{d|k}\mu(d)\frac{k}{d} =$$

$$\sum_{k=1}^{n}\sum_{dd'=k} d'\mu(d) = \sum_{d=1}^{n}\mu(d)\sum_{d'\leq\frac{n}{d}} d'.$$

Since $\sum_{j=1}^{m} j = \frac{1}{2}m(m+1)$, we have

$$\sum_{k=1}^{n}\phi(k) = \sum_{d=1}^{n}\mu(d)\sum_{d'\leq\frac{n}{d}} d' = \frac{1}{2}\sum_{d=1}^{n}\mu(d)[\frac{n}{d}]([\frac{n}{d}]+1). \tag{2.8}$$

We have $[\frac{n}{d}]([\frac{n}{d}]+1) \leq \frac{n}{d}(\frac{n}{d}+1) = (\frac{n}{d})^2 + \frac{n}{d}$, and $[\frac{n}{d}]([\frac{n}{d}]+1) \geq (\frac{n}{d}-1)\frac{n}{d} = (\frac{n}{d})^2 - \frac{n}{d}$; thus,

$$(\frac{n}{d})^2 - \frac{n}{d} \leq [\frac{n}{d}]([\frac{n}{d}]+1) \leq (\frac{n}{d})^2 + \frac{n}{d}.$$

Substituting this two-sided inequality in (2.8), we obtain

$$\frac{n^2}{2}\sum_{d=1}^{n}\frac{\mu(d)}{d^2} - \frac{n}{2}\sum_{d=1}^{n}\frac{\mu(d)}{d} \leq \sum_{k=1}^{n}\phi(k) \leq \frac{n^2}{2}\sum_{d=1}^{n}\frac{\mu(d)}{d^2} + \frac{n}{2}\sum_{d=1}^{n}\frac{\mu(d)}{d}. \tag{2.9}$$

Now

$$|\sum_{d=1}^{n}\frac{\mu(d)}{d}|\le\sum_{d=1}^{n}\frac{1}{d}=1+\sum_{d=2}^{n}\frac{1}{d}\le 1+\log n, \tag{2.10}$$

since the final sum is a lower Riemann sum for $\int_1^n \frac{1}{x}\,dx$. From (2.9) and (2.10), we obtain

$$\lim_{n\to\infty}\frac{\sum_{k=2}^{n}\phi(k)}{n(n-1)}=\frac{1}{2}\sum_{d=1}^{\infty}\frac{\mu(d)}{d^2}. \tag{2.11}$$

It remains to evaluate $\sum_{d=1}^{\infty}\frac{\mu(d)}{d^2}$. On the face of it, from the definition of μ, it would seem very difficult to evaluate this explicitly. However, Möbius inversion saves the day. Consider (2.2)–(2.4) with $a = 1$ and $b = \mu$ and with $x = 2$. With these choices, the right hand sides of (2.2) and (2.3) are absolutely convergent. By (2.5), we have $1 * \mu = e$; that is, $a * b = e$. Therefore, we conclude from (2.2)–(2.4) that

$$\left(\sum_{d=1}^{\infty}\frac{1}{d^2}\right)\left(\sum_{d=1}^{\infty}\frac{\mu(d)}{d^2}\right)=1. \tag{2.12}$$

Recall the well-known formula

$$\sum_{n=1}^{\infty}\frac{1}{n^2}=\frac{\pi^2}{6}. \tag{2.13}$$

We give a completely elementary proof of this fact in Appendix D. From (2.12) and (2.13) we obtain

$$\sum_{d=1}^{\infty}\frac{\mu(d)}{d^2}=\frac{6}{\pi^2}. \tag{2.14}$$

Using (2.14) with (2.11) and (2.6) gives

$$\lim_{n\to\infty}q_n=\frac{6}{\pi^2},$$

completing the proof of the theorem. □

Remark. If a is an arithmetic function and f is a nondecreasing function, we say that the function f is the *average order* of the arithmetic function a if $\frac{1}{n}\sum_{k=1}^{n}a(k) = f(n)+o(f(n))$. Of course this doesn't uniquely define f; we usually choose a particular such f which has a simple form. From (2.11) and (2.14), it follows that the average order of the Euler ϕ-function is $\frac{3n}{\pi^2}$.

We now turn to Theorem 2.2.

Proof of Theorem 2.2. From the definition of the Möbius function, it follows that

$$\mu^2(n) = \begin{cases} 1, & \text{if } n \text{ is square-free;} \\ 0, & \text{otherwise.} \end{cases} \tag{2.15}$$

Thus, letting

$$A_n = \{j \in [n] : j \text{ is square-free}\},$$

we have

$$|A_n| = \sum_{j=1}^{n} \mu^2(j). \tag{2.16}$$

To prove the theorem, we need to show that

$$\lim_{n\to\infty} \frac{|A_n|}{n} = \frac{6}{\pi^2}. \tag{2.17}$$

We need the following lemma.

Lemma 2.2.

$$\mu^2(n) = \sum_{k^2|n} \mu(k).$$

Proof. Let $\Lambda(n) := \sum_{k^2|n} \mu(k)$. If n is square-free, then the only integer k that satisfies $k^2|n$ is $k = 1$. Thus, since $\mu(1) = 1$, we have $\Lambda(n) = 1$. On the other hand, if n is not square-free, then n can be written in the form $n = m^2 l$, where $m > 1$ and l is square-free. Now $k^2|m^2 l$ if and only if $k|m$. (The reader should verify this.) Thus, we have

$$\Lambda(n) = \sum_{k^2|n} \mu(k) = \sum_{k^2|m^2 l} \mu(k) = \sum_{k|m} \mu(k) = (\mu * 1)(m) = 0,$$

where the last equality follows from (2.5). The lemma now follows from (2.15). □

Using Lemma 2.2, we have

$$\sum_{j=1}^{n} \mu^2(j) = \sum_{j=1}^{n} \sum_{k^2|j} \mu(k). \tag{2.18}$$

If $k^2 > n$, then $\mu(k)$ will not appear on the right hand side of (2.18). If $k^2 \le n$, then $\mu(k)$ will appear on the right hand side of (2.18) $[\frac{n}{k^2}]$ times, namely, when $j = k^2, 2k^2, \ldots, [\frac{n}{k^2}]k^2$. Thus, we have

$$\sum_{j=1}^{n} \mu^2(j) = \sum_{j=1}^{n} \sum_{k^2|j} \mu(k) = \sum_{k^2 \le n} [\frac{n}{k^2}]\, \mu(k) = \sum_{k \le [n^{\frac{1}{2}}]} [\frac{n}{k^2}]\, \mu(k) =$$

$$n \sum_{k \le [n^{\frac{1}{2}}]} \frac{\mu(k)}{k^2} + \sum_{k \le [n^{\frac{1}{2}}]} \Big([\frac{n}{k^2}] - \frac{n}{k^2}\Big)\mu(k). \tag{2.19}$$

Since each summand in the second term on the right hand side of (2.19) is bounded in absolute value by 1, we have

$$|\sum_{k \le [n^{\frac{1}{2}}]} \Big([\frac{n}{k^2}] - \frac{n}{k^2}\Big)\mu(k)| \le n^{\frac{1}{2}}. \tag{2.20}$$

It follows from (2.16), (2.19), and (2.20) that

$$\lim_{n \to \infty} \frac{|A_n|}{n} = \sum_{k=1}^{\infty} \frac{\mu(k)}{k^2}.$$

Using this with (2.14) gives (2.17) and completes the proof of the theorem. □

We now give a heuristic probabilistic proof and a rigorous probabilistic proof of Theorem 2.1. In the heuristic proof, we put quotation marks around the steps that are not rigorous.

Heuristic Probabilistic Proof of Theorem 2.1. Let $\{p_k\}_{k=1}^{\infty}$ be an enumeration of the primes. In the spirit described in the first paragraph of the chapter, if we pick a positive integer "at random," then the "probability" of it being divisible by the prime number p_k is $\frac{1}{p_k}$. (Of course, this is true also with p_k replaced by an arbitrary positive integer.) If we pick two positive integers "independently," then the "probability" that they are both divisible by p_k is $\frac{1}{p_k}\frac{1}{p_k} = \frac{1}{p_k^2}$, by "independence." So the "probability" that at least one of them is not divisible by p_k is $1 - \frac{1}{p_k^2}$. The "probability" that a "randomly" selected positive integer is divisible by the two distinct primes, p_j and p_k, is $\frac{1}{p_j p_k} = \frac{1}{p_j}\frac{1}{p_k}$. (The reader should check that this "holds" more generally if p_j and p_k are replaced by an arbitrary pair of relatively prime positive integers, but not otherwise.) Thus, the events of *being divisible by* p_j and *being divisible by* p_k are "independent." Now two "randomly" selected positive integers are relatively prime if and only if, for every k, at least one of the integers is not divisible by p_k. But since the "probability" that at least one of them is not divisible by p_k is $1-\frac{1}{p_k^2}$, and since being divisible by a prime p_j and being divisible by a different prime p_k are "independent" events, the "probability" that the two "randomly" selected positive integers are such that, for every k, at least one of them is not divisible by p_k is $\prod_{k=1}^{\infty}(1-\frac{1}{p_k^2})$. Thus, this should be the "probability" that two "randomly" selected positive integers are relatively prime. □

Rigorous Probabilistic Proof of Theorem 2.1. For the probabilistic proof, the second alternative suggested in the second paragraph of the chapter will be more convenient. Thus, we choose an integer from $[n]$ uniformly at random and then choose a second integer from $[n]$ uniformly at random. Let $\Omega_n = [n]$. The appropriate probability space on which to analyze the model described above is the space $(\Omega_n \times \Omega_n, P_n)$, where the probability measure P_n on $\Omega_n \times \Omega_n$ is the uniform measure; that is, $P_n(A) = \frac{|A|}{n^2}$, for any $A \subset \Omega_n \times \Omega_n$. The point $(i, j) \in \Omega_n \times \Omega_n$ indicates that the integer i was chosen the first time and the integer j was chosen the second time. Let C_n denote the event that the two selected integers are relatively prime; that is,

$$C_n = \{(i, j) \in \Omega_n \times \Omega_n : \gcd(i, j) = 1\}.$$

Then the probability q_n that the two selected integers are relatively prime is

$$q_n = P_n(C_n) = \frac{|C_n|}{n^2}.$$

Let $\{p_k\}_{k=1}^{\infty}$ denote the prime numbers arranged in increasing order. (Any enumeration of the primes would do, but for the proof it is more convenient to choose the increasing enumeration.) For each $k \in \mathbb{N}$, let $B_{n;k}^1$ denote the event that the first integer chosen is divisible by p_k and let $B_{n;k}^2$ denote the event that the second integer chosen is divisible by p_k. That is,

$$B_{n;k}^1 = \{(i, j) \in \Omega_n \times \Omega_n : p_k|i\}, \quad B_{n;k}^2 = \{(i, j) \in \Omega_n \times \Omega_n : p_k|j\}.$$

Note of course that the above sets are empty if $p_k > n$. The event $B_{n;k}^1 \cap B_{n;k}^2 = \{(i, j) \in \Omega_n \times \Omega_n : p_k|i \text{ and } p_k|j\}$ is the event that both selected integers have p_k as a factor. There are $[\frac{n}{p_k}]$ integers in Ω_n that are divisible by p_k, namely, $p_k, 2p_k, \cdots, [\frac{n}{p_k}]p_k$. Thus, there are $[\frac{n}{p_k}]^2$ pairs $(i, j) \in \Omega_n \times \Omega_n$ for which both coordinates are divisible by p_k; therefore,

$$P_n(B_{n;k}^1 \cap B_{n;k}^2) = \frac{[\frac{n}{p_k}]^2}{n^2}. \tag{2.21}$$

Note that $\cup_{k=1}^{\infty}(B_{n;k}^1 \cap B_{n;k}^2) = \cup_{k=1}^{n}(B_{n;k}^1 \cap B_{n;k}^2)$ is the event that the two selected integers have at least one common prime factor. (The equality above follows from the fact that $B_{n;k}^1$ and $B_{n;k}^2$ are clearly empty for $k > n$.) Consequently, C_n can be expressed as

$$C_n = \left(\cup_{k=1}^n (B_{n;k}^1 \cap B_{n;k}^2)\right)^c = \cap_{k=1}^n (B_{n;k}^1 \cap B_{n;k}^2)^c,$$

where $A^c := \Omega_n \times \Omega_n - A$ denotes the complement of an event $A \subset \Omega_n \times \Omega_n$. Thus,

$$P_n(C_n) = P_n\left(\cap_{k=1}^n (B_{n;k}^1 \cap B_{n;k}^2)^c\right). \tag{2.22}$$

Let $R < n$ be a positive integer. We have

$$\cap_{k=1}^n (B_{n;k}^1 \cap B_{n;k}^2)^c = \cap_{k=1}^R (B_{n;k}^1 \cap B_{n;k}^2)^c - \cup_{k=R+1}^n (B_{n;k}^1 \cap B_{n;k}^2)$$

and, of course, $\cap_{k=1}^n (B_{n;k}^1 \cap B_{n;k}^2)^c \subset \cap_{k=1}^R (B_{n;k}^1 \cap B_{n;k}^2)^c$. Thus,

$$P_n\left(\cap_{k=1}^R (B_{n;k}^1 \cap B_{n;k}^2)^c\right) - P_n\left(\cup_{k=R+1}^n (B_{n;k}^1 \cap B_{n;k}^2)\right) \le$$
$$P_n\left(\cap_{k=1}^n (B_{n;k}^1 \cap B_{n;k}^2)^c\right) \le P_n\left(\cap_{k=1}^R (B_{n;k}^1 \cap B_{n;k}^2)^c\right). \tag{2.23}$$

Using the sub-additivity property of probability measures for the first inequality below, and using (2.21) for the equality below, we have

$$P_n\left(\cup_{k=R+1}^n (B_{n;k}^1 \cap B_{n;k}^2)\right) \le \sum_{k=R+1}^n P_n(B_{n;k}^1 \cap B_{n;k}^2) = \sum_{k=R+1}^n \frac{[\frac{n}{p_k}]^2}{n^2} \le \sum_{k=R+1}^\infty \frac{1}{p_k^2}. \tag{2.24}$$

Up until now, we have made no assumption on n. Now assume that $p_k | n$, for $k = 1, \cdots, R$; that is, assume that n is a multiple of $\prod_{k=1}^R p_k$. Denote the set of such n by D_R; that is,

$$D_R = \{n \in \mathbb{N} : p_k | n \text{ for } k = 1, \cdots, R\}.$$

Recall that the event $B_{n;k}^1 \cap B_{n;k}^2$ is the event that both selected integers are divisible by k. We claim that if $n \in D_R$, then the events $\{B_{n;k}^1 \cap B_{n;k}^2\}_{k=1}^R$ are independent. That is, for any subset $I \subset \{1, 2, \cdots, R\}$, one has

$$P_n\left(\cap_{k\in I} (B_{n;k}^1 \cap B_{n;k}^2)\right) = \prod_{k\in I} P_n(B_{n;k}^1 \cap B_{n;k}^2), \text{ if } n \in D_R. \tag{2.25}$$

The proof of (2.25) is a straightforward counting exercise and is left as Exercise 2.4. If events $\{A_k\}_{k=1}^R$ are independent, then the complementary events $\{A_k^c\}_{k=1}^R$ are also independent. See Exercise A.3 in Appendix A. Thus, we conclude that

$$P_n\left(\cap_{k=1}^R (B_{n;k}^1 \cap B_{n;k}^2)^c\right) = \prod_{k=1}^R P_n\left((B_{n;k}^1 \cap B_{n;k}^2)^c\right), \text{ if } n \in D_R. \tag{2.26}$$

By (2.21) we have $P_n\big((B^1_{n;k} \cap B^2_{n;k})^c\big) = 1 - P_n(B^1_{n;k} \cap B^2_{n;k}) = 1 - \frac{[\frac{n}{p_k}]^2}{n^2}$, for any n. Thus, from the definition of D_R, we have

$$P_n\big((B^1_{n;k} \cap B^2_{n;k})^c\big) = 1 - \frac{1}{p_k^2}, \text{ if } n \in D_R. \tag{2.27}$$

From (2.22) to (2.24), (2.26), and (2.27), we conclude that

$$\prod_{k=1}^{R}(1-\frac{1}{p_k^2}) - \sum_{k=R+1}^{\infty}\frac{1}{p_k^2} \le P_n(C_n) \le \prod_{k=1}^{R}(1-\frac{1}{p_k^2}), \text{ for } R \in \mathbb{N} \text{ and } n \in D_R. \tag{2.28}$$

We now use (2.28) to obtain an estimate on $P_n(C_n)$ for general n. Let $n \ge \prod_{k=1}^R p_k$. Let n' denote the largest integer in D_R which is smaller or equal to n, and let n'' denote the smallest integer in D_R which is larger or equal to n. Since D_R is the set of positive multiples of $\prod_{k=1}^R p_k$, we obviously have

$$n' > n - \prod_{k=1}^{R} p_k \text{ and } n'' < n + \prod_{k=1}^{R} p_k. \tag{2.29}$$

For any n, note that $n^2 P_n(C_n) = |C_n|$ is the number of pairs $(i, j) \in \Omega_n \times \Omega_n$ that are relatively prime. Obviously, the number of such pairs is increasing in n. Thus $(n')^2 P_{n'}(C_{n'}) \le n^2 P_n(C_n) \le (n'')^2 P_{n''}(C_{n''})$, or equivalently,

$$(\frac{n'}{n})^2 P_{n'}(C_{n'}) \le P_n(C_n) \le (\frac{n''}{n})^2 P_{n''}(C_{n''}). \tag{2.30}$$

Since $n', n'' \in D_R$, we conclude from (2.28)–(2.30) that

$$(\frac{n-\prod_{k=1}^R p_k}{n})^2(\prod_{k=1}^{R}(1-\frac{1}{p_k^2}) - \sum_{k=R+1}^{\infty}\frac{1}{p_k^2}) < P_n(C_n) < (\frac{n+\prod_{k=1}^R p_k}{n})^2\prod_{k=1}^{R}(1-\frac{1}{p_k^2}). \tag{2.31}$$

Letting $n \to \infty$ in (2.31), we obtain

$$\prod_{k=1}^{R}(1-\frac{1}{p_k^2}) - \sum_{k=R+1}^{\infty}\frac{1}{p_k^2} \le \liminf_{n\to\infty} P_n(C_n) \le \limsup_{n\to\infty} P_n(C_n) \le \prod_{k=1}^{R}(1-\frac{1}{p_k^2}). \tag{2.32}$$

Now (2.32) holds for arbitrary R; thus letting $R \to \infty$, we conclude that

$$\lim_{n\to\infty} P_n(C_n) = \prod_{k=1}^{\infty}(1-\frac{1}{p_k^2}). \tag{2.33}$$

The celebrated *Euler product formula* states that

$$\frac{1}{\prod_{k=1}^{\infty}(1-\frac{1}{p_k^r})}=\sum_{n=1}^{\infty}\frac{1}{n^r},\ r>1; \tag{2.34}$$

see Exercise 2.5. From (2.33), (2.34), and (2.13), we conclude that

$$\lim_{n\to\infty} q_n=\lim_{n\to\infty} P_n(C_n)=\frac{1}{\sum_{n=1}^{\infty}\frac{1}{n^2}}=\frac{6}{\pi^2}. \qquad \square$$

Exercise 2.1. Give a direct proof of Corollary 2.1. (Hint: The Euler ϕ-function $\phi(n)$ counts the number of positive integers that are less than or equal to n and relatively prime to n. We employ the *sieve method*, which from the point of view of set theory is the method of *inclusion–exclusion*. Start with a list of all n integers between 1 and n as potential members of the set of the $\phi(n)$ relatively prime integers to n. Let $\{p_j\}_{j=1}^m$ be the prime divisors of n. For any such p_j, the $\frac{n}{p_j}$ numbers $p_j, 2p_j, \ldots, \frac{n}{p_j}p_j$ are not relatively prime to n. So we should strike these numbers from our list. When we do this for each j, the remaining numbers on the list are those numbers that are relatively prime to n, and the size of the list is $\phi(n)$. Now we haven't necessarily reduced the size of our list to $N_1 := n-\sum_{j=1}^m \frac{n}{p_j}$, because some of the numbers we have deleted might be multiples of two different primes, p_i and p_j, in which case they were subtracted above twice. Thus we need to add back to N_1 all of the $\frac{n}{p_i p_j}$ multiples of $p_i p_j$, for $i \neq j$. That is, we now have $N_2 := N_1+\sum_{i\neq j}\frac{n}{p_i p_j}$. Continue in this vein.

Exercise 2.2. This exercise presents an alternative proof to Proposition 2.2:

(a) Show that the arithmetic function $\sum_{d|n}\phi(d)$ is multiplicative. Use the fact that ϕ is multiplicative—see Exercise 2.3.
(b) Show that $\sum_{d|n}\phi(d)=n$, when n is a prime power.
(c) Conclude that Proposition 2.2 holds.

Exercise 2.3. The *Chinese remainder theorem* states that if n and m are relatively prime positive integers, and $a\in[n]$ and $b\in[m]$, then there exists a unique $c\in[nm]$ such that $c=a \bmod n$ and $c=b \bmod m$. (For a proof, see [27].) Use this to prove that the Euler ϕ-function is multiplicative. Then use the fact that ϕ is multiplicative to prove (2.7).

Exercise 2.4. Prove (2.25).

Exercise 2.5. Prove the Euler product formula (2.34). (Hint: Let N_ℓ denote the set of positive integers all of whose prime factors are in the set $\{p_k\}_{k=1}^{\ell}$. Using the fact that

$$\frac{1}{1-\frac{1}{p_k^r}}=\sum_{m=0}^{\infty}\frac{1}{p_k^{rm}},$$

for all $k \in \mathbb{N}$, first show that $\frac{1}{1-\frac{1}{p_1^r}}\frac{1}{1-\frac{1}{p_2^r}} = \sum_{n\in N_2}\frac{1}{n^r}$, and then show that $\prod_{k=1}^{\ell}\frac{1}{1-\frac{1}{p_k^r}} = \sum_{n\in N_\ell}\frac{1}{n^r}$, for any $\ell \in \mathbb{N}$.)

Exercise 2.6. Using Theorem 2.1, prove the following result: Let $2 \le d \in \mathbb{N}$. Choose two integers uniformly at random from $[n]$. As $n \to \infty$, the asymptotic probability that their greatest common divisor is d is $\frac{6}{d^2\pi^2}$.

Exercise 2.7. Give a probabilistic proof of Theorem 2.2.

Chapter Notes

It seems that Theorem 2.1 was first proven by E. Cesàro in 1881. A good source for the results in this chapter is Nathanson's book [27]. See also the more advanced treatment of Tenenbaum [33], which contains many interesting and nontrivial exercises. The heuristic probabilistic proof of Theorem 2.1 is well known and can be found readily, including via a Google-search. I am unaware of a rigorous probabilistic proof in the literature.

Chapter 3
A One-Dimensional Probabilistic Packing Problem

Consider n molecules lined up in a row. From among the $n-1$ nearest neighbor pairs, select one pair at random and "bond" the two molecules together. Now from all the remaining nearest neighbor pairs, select one pair at random and bond the two molecules together. Continue like this until no nearest neighbor pairs remain. Let $M_{n;2}$ denote the random variable that counts the number of bonded molecules. Let $EM_{n;2}$ denote the expected value of $M_{n;2}$, that is, the average number of bonded molecules. The first thing we would like to do is to compute the limiting average fraction of bonded molecules: $\lim_{n\to\infty}\frac{EM_{n;2}}{n}$. Then we would like to show that $\frac{M_{n;2}}{n}$ is close to this limiting average with high probability as $n\to\infty$; that is, we would like to prove that $\frac{M_{n;2}}{n}$ satisfies the *weak law of large numbers*.

Of course, by definition, $EM_{n;2}=\sum_{j=0}^{n} jP(M_{n;2}=j)$, where $P(M_{n;2}=j)$ is the probability that $M_{n;2}$ is equal to j. However, it would be fruitless to pursue this formula to evaluate $EM_{n;2}$ asymptotically because the calculation of $P(M_{n;2}=j)$ is hopelessly complicated. We will solve the problem with the help of generating functions.

Actually, we will consider a slightly more general problem, where the pairs are replaced by k-tuples, for some $k\ge 2$. So the problem is as follows. There are n molecules on a line. From among the $n-k+1$ nearest neighbor k-tuples, select one at random and "bond" the k molecules together. Now from among all the remaining nearest neighbor k-tuples, select one at random and bond the k molecules together. Continue like this until there are no nearest neighbor k-tuples left. Let $M_{n;k}$ denote the random variable that counts the number of bonded molecules, and let $EM_{n;k}$ denote the expected value of $M_{n;k}$. See Fig. 3.1. Here is our result.

Theorem 3.1. *For each integer* $k\ge 2$,

$$\lim_{n\to\infty}\frac{EM_{n;k}}{n}=k\exp(-2\sum_{j=1}^{k-1}\frac{1}{j})\int_0^1\exp(2\sum_{j=1}^{k-1}\frac{s^j}{j})\,ds:=p_k. \tag{3.1}$$

Furthermore, $\frac{M_{n;k}}{n}$ *satisfies the weak law of large numbers; that is, for all* $\epsilon>0$,

R.G. Pinsky, *Problems from the Discrete to the Continuous*, Universitext,
DOI 10.1007/978-3-319-07965-3_3, © Springer International Publishing Switzerland 2014

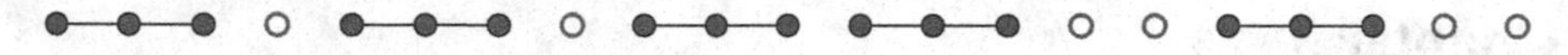

Fig. 3.1 A realization with n = 21 and k = 3 that gives $M_{21,3} = 15$

$$\lim_{n\to\infty} P(|\frac{M_{n;k}}{n} - p_k| \geq \epsilon) = 0. \tag{3.2}$$

Remark 1. Only when $k = 2$ can p_k be calculated explicitly; one obtains $p_2 = 1-e^{-2} \approx 0.865$. Numerical integration gives $p_3 \approx 0.824$, $p_4 \approx 0.804$, $p_5 \approx 0.792$, $p_{10} \approx 0.770$, $p_{100} \approx 0.750$, $p_{1000} \approx 0.748$, and $p_{10,000} = 0.748$. The expression p_k seems surprisingly difficult to analyze. We suggest the following open problem to the reader.

Open Problem. Prove that p_k is monotone decreasing and calculate $\lim_{k\to\infty} p_k$.

Remark 2. Any molecule that remains unbonded at the end of the nearest neighbor k-tuple bonding process occurs in a maximal row of j unbonded molecules, for some $j \in [k-1]$. In the limit as $n \to \infty$, what fraction of molecules ends up in a maximal row of j unbounded molecules? See Exercise 3.2. (In Fig. 3.1, numbering from left to right, molecules #4 and #8 occur in a maximal row of one unbounded molecule, while molecules #15, #16, #20, and #21 occur in a maximal row of two unbounded molecules.)

Proof. For notational convenience, let $H_n^{(k)} = EM_{n;k}$ and $L_n^{(k)} = EM_{n;k}^2$. To prove the theorem, it suffices to show that

$$EM_{n;k} = H_n^{(k)} = p_k n + o(n), \text{ as } n \to \infty, \tag{3.3}$$

and that

$$EM_{n;k}^2 = L_n^{(k)} = p_k^2 n^2 + o(n^2), \text{ as } n \to \infty. \tag{3.4}$$

This method of proof is known as the *second moment method.* It is clear that (3.1) follows from (3.3). An application of Chebyshev's inequality shows that (3.2) follows from (3.3) and (3.4). To see this, note that if Z is a random variable with expected value EZ and variance $\sigma^2(Z)$, then Chebyshev's inequality states that

$$P(|Z - EZ| \geq \delta) \leq \frac{\sigma^2(Z)}{\delta^2}, \text{ for any } \delta > 0.$$

Also, $\sigma^2(Z) = EZ^2 - (EZ)^2$. We apply Chebyshev's inequality with $Z = \frac{M_{n;k}}{n}$. Using (3.3) and (3.4), we have

$$EZ = \frac{H_n^{(k)}}{n} = p_k + o(1), \text{ as } n \to \infty, \tag{3.5}$$

and

$$\sigma^2(Z) = \frac{L_n^{(k)}}{n^2} - \frac{(H_n^{(k)})^2}{n^2} = p_k^2 + o(1) - (p_k + o(1))^2 = o(1), \text{ as } n \to \infty.$$

Thus, we obtain for all $\delta > 0$,

$$P(|\frac{M_{n;k}}{n} - \frac{H_n^{(k)}}{n}| \geq \delta) \leq \frac{o(1)}{\delta^2}, \text{ as } n \to \infty,$$

or, equivalently,

$$\lim_{n\to\infty} P(|\frac{M_{n;k}}{n} - \frac{H_n^{(k)}}{n}| \geq \delta) = 0, \text{ for all } \delta > 0. \tag{3.6}$$

We now show that (3.2) follows from (3.3) and (3.6). Fix $\epsilon > 0$. We have

$$|\frac{M_{n;k}}{n} - p_k| = |\frac{M_{n;k}}{n} - \frac{H_n^{(k)}}{n} + \frac{H_n^{(k)}}{n} - p_k| \leq |\frac{M_{n;k}}{n} - \frac{H_n^{(k)}}{n}| + |\frac{H_n^{(k)}}{n} - p_k|.$$

For sufficiently large n_ϵ, one has from (3.3) that $|\frac{H_n^{(k)}}{n} - p_k| \leq \frac{\epsilon}{2}$, for $n \geq n_\epsilon$. Thus, for $n \geq n_\epsilon$, a necessary condition for $|\frac{M_{n;k}}{n} - p_k| \geq \epsilon$ is that $|\frac{M_{n;k}}{n} - \frac{H_n^{(k)}}{n}| \geq \frac{\epsilon}{2}$. Consequently,

$$P(|\frac{M_{n;k}}{n} - p_k| \geq \epsilon) \leq P(|\frac{M_{n;k}}{n} - \frac{H_n^{(k)}}{n}| \geq \frac{\epsilon}{2}), \text{ for } n \geq n_\epsilon.$$

Now (3.2) follows from this and (3.6).

Our proofs of (3.3) and (3.4) will follow similar lines. Before commencing with the proof of (3.3), we trace its general architecture. Only the first step of the proof involves probability. In this step, we employ probabilistic reasoning to produce a recursion equation that gives $H_n^{(k)}$ in terms of $H_0^{(k)}, H_1^{(k)}, \ldots, H_{n-k}^{(k)}$. In this form, the equation is not useful because as $n \to \infty$, it gives $H_n^{(k)}$ in terms of a growing number of its predecessors. However, defining $S_n^{(k)} = \sum_{j=0}^{n} H_j^{(k)}$, and using the abovementioned recursion equation, we find that $S_n^{(k)}$ is given in terms of only two of its predecessors. We then construct the generating function $g(t)$ whose coefficients are $\{S_n^{(k)}\}_{n=0}^{\infty}$. Using the recursion equation for $S_n^{(k)}$, we show that g solves a linear, first order differential equation. We solve this differential equation to obtain an explicit formula for $g(t)$. This explicit formula reveals that g possesses a singularity at $t = 1$. Exploiting this singularity allows us to evaluate $\lim_{n\to\infty} \frac{S_n^{(k)}}{n^2}$, and then a simple observation allows us to obtain $\lim_{n\to\infty} \frac{H_n^{(k)}}{n}$ from $\lim_{n\to\infty} \frac{S_n^{(k)}}{n^2}$.

We now commence with the proof of (3.3). Note that if we start with $n < k$ molecules, then none of them will get bonded. Thus,

$$H_n^{(k)} = 0, \text{ for } n = 0, \ldots, k-1. \tag{3.7}$$

We now derive a recursion relation for $H_n^{(k)}$. The method we use is called *first step analysis*. We begin with a line of $n \geq k$ unbonded molecules, and in the first step, one of the nearest neighbor k-tuples is chosen at random and its k molecules are bonded. In order from left to right, denote the original $n-k+1$ nearest neighbor k-tuples by $\{B_j\}_{j=1}^{n-k+1}$. If B_j was chosen in the first step, then the original row now contains a row of $j-1$ unbonded molecules to the left of the bonded k-tuple B_j and a row of $n+1-j-k$ unbonded molecules to the right of B_j. To complete the random bonding process, we choose random k-tuples from these two sub-rows until there are no more k-tuples to choose from. This gives us the following formula for the conditional expectation of $M_{n;k}$ given that B_j was selected first: for $n \geq k$,

$$E(M_{n;k}|B_j \text{ selected first}) = k + E(M_{j-1;k} + M_{n+1-j-k;k}) = k + H_{j-1}^{(k)} + H_{n+1-j-k}^{(k)}. \tag{3.8}$$

Of course, for each $j \in [n-k+1]$, the probability that B_j was chosen first is $\frac{1}{n-k+1}$. Thus, we obtain the formula

$$EM_{n;k} = H_n^{(k)} = \sum_{j=1}^{n-k+1} P(B_j \text{ selected first}) E(M_{n;k}|B_j \text{ selected first}) =$$

$$\frac{1}{n-k+1} \sum_{j=1}^{n-k+1} (k + H_{j-1}^{(k)} + H_{n+1-j-k}^{(k)}), \ n \geq k.$$

We can rewrite this as

$$H_n^{(k)} = k + \frac{2}{n-k+1} \sum_{j=0}^{n-k} H_j^{(k)}, \ n \geq k. \tag{3.9}$$

The above recursion equation is not useful directly because it gives $H_n^{(k)}$ in terms of $n-k+1$ of its predecessors; we want a recursion equation that expresses a given term in terms of a fixed finite number of its predecessors. To that end, we define

$$S_n^{(k)} = \sum_{j=0}^{n} H_j^{(k)}. \tag{3.10}$$

Substituting this in (3.9) gives

$$H_n^{(k)} = k + \frac{2}{n-k+1} S_{n-k}^{(k)}, \ n \geq k. \tag{3.11}$$

Writing (3.7) and (3.11) in terms of $\{S_n^{(k)}\}_{n=0}^{\infty}$, we obtain

$$S_n^{(k)} = 0, \text{ for } n = 0, \ldots, k-1, \tag{3.12}$$

and

$$S_n^{(k)} - S_{n-1}^{(k)} = k + \frac{2}{n-k+1} S_{n-k}^{(k)}, \ n \geq k. \tag{3.13}$$

This recursion equation has the potential to be useful since it gives $S_n^{(k)}$ in terms of only two of its predecessors—$S_{n-1}^{(k)}$ and $S_{n-k}^{(k)}$. Of course, we have paid a price—we are now working with $S_n^{(k)}$ instead of $H_n^{(k)}$; but this will be dealt with easily. For convenience, we drop the superscript k from $S_n^{(k)}$, $H_n^{(k)}$, and $L_n^{(k)}$ for the rest of the chapter, except in the statement of propositions. We rewrite (3.13) as

$$(n-k+1)S_n = (n-k+1)S_{n-1} + 2S_{n-k} + k(n-k+1), \ n \geq k. \tag{3.14}$$

We now define the generating function for $\{S_n\}_{n=0}^{\infty}$ and use (3.14) to derive a linear, first-order differential equation that is satisfied by this generating function. The generating function $g(t)$ is defined by

$$g(t) = \sum_{n=0}^{\infty} S_n t^n = \sum_{n=k}^{\infty} S_n t^n, \tag{3.15}$$

where the second equality follows from (3.12). From the definitions, it follows that $H_n \leq n$, and thus $S_n \leq \frac{1}{2}n(n+1)$. Consequently, the sum on the right hand side of (3.15) converges for $|t| < 1$, with the convergence being uniform for $|t| \leq \rho$, for any $\rho \in (0, 1)$. It follows then that

$$g'(t) = \sum_{n=k}^{\infty} n S_n t^{n-1}, \ |t| < 1. \tag{3.16}$$

Multiply equation (3.14) by t^n and group the terms in the following way:

$$nS_n t^n - (k-1)S_n t^n = (n-1)S_{n-1}t^n - (k-2)S_{n-1}t^n + 2S_{n-k}t^n + k(n-k+1)t^n.$$

Now summing the equation over all $n \geq k$, and appealing to (3.15), (3.16), and (3.12), we obtain the differential equation

$$tg'(t) - (k-1)g(t) = t^2 g'(t) - (k-2)tg(t) + 2t^k g(t) + kt \sum_{n=k}^{\infty} nt^{n-1} - k(k-1) \sum_{n=k}^{\infty} t^n. \tag{3.17}$$

Since $\sum_{n=k}^{\infty} nt^{n-1}$ is the derivative of $\sum_{n=k}^{\infty} t^n = \frac{t^k}{1-t}$, it follows that $\sum_{n=k}^{\infty} nt^{n-1} = (\frac{t^k}{1-t})' = \frac{(1-t)kt^{k-1}+t^k}{(1-t)^2}$. Using these facts and doing some algebra, which leads to many cancelations, we obtain

$$kt\sum_{n=k}^{\infty} nt^{n-1} - k(k-1)\sum_{n=k}^{\infty} t^n = \frac{kt^k}{(1-t)^2}. \tag{3.18}$$

Substituting this in (3.17), and doing a little algebra, we obtain

$$g'(t) = \frac{(k-1)-(k-2)t+2t^k}{t(1-t)}g(t) + \frac{kt^{k-1}}{(1-t)^3},\ 0 < t < 1. \tag{3.19}$$

Note that we have excluded $t = 0$ because we have divided by t.

There are two singularities in the above equation—one at $t = 0$ and one at $t = 1$. The singularity at $t = 0$ is removable; indeed, $g(0) = 0$ so the first term on the right hand side of (3.19) can be defined at 0. The singularity at 1, on the other hand, is authentic, and actually contains the solution to our problem—we will just need to "unzip" it.

The linear, first-order differential equation in (3.19) is written in the form $g'(t) = a(t)g(t) + b(t)$, where

$$a(t) = \frac{(k-1)-(k-2)t+2t^k}{t(1-t)},\quad b(t) = \frac{kt^{k-1}}{(1-t)^3}. \tag{3.20}$$

Let $\epsilon \in (0,1)$ and rewrite the differential equation as

$$(g(t)e^{-\int_\epsilon^t a(s)\,ds})' = b(t)e^{-\int_\epsilon^t a(s)\,ds}.$$

Integrating from ϵ to $t \in (\epsilon, 1)$ gives

$$g(t)e^{-\int_\epsilon^t a(r)\,dr} = g(\epsilon) + \int_\epsilon^t b(s)e^{-\int_\epsilon^s a(r)\,dr}\,ds,\ t \in (\epsilon, 1),$$

which we rewrite as

$$g(t) = g(\epsilon)e^{\int_\epsilon^t a(r)\,dr} + \int_\epsilon^t b(s)e^{\int_s^t a(r)\,dr}\,ds,\ t \in (\epsilon, 1). \tag{3.21}$$

Since $\lim_{t\to 0} ta(t) = k-1$, there exists a $t_0 > 0$ such that $a(t) \le \frac{k-\frac{1}{2}}{t}$, for $0 < t \le t_0$. Thus, for $\epsilon < t_0$, one has

$$e^{\int_\epsilon^{t_0} a(r)\,dr} \le e^{\int_\epsilon^{t_0} \frac{k-\frac{1}{2}}{r}\,dr} = (\frac{t_0}{\epsilon})^{k-\frac{1}{2}}.$$

By (3.15) we have $g(\epsilon) = O(\epsilon^k)$ as $\epsilon \to 0$. Therefore,

$$\lim_{\epsilon\to 0} g(\epsilon)e^{\int_\epsilon^t a(r)\,dr} = \lim_{\epsilon\to 0} g(\epsilon)e^{\int_\epsilon^{t_0} a(r)\,dr}e^{\int_{t_0}^t a(r)\,dr} \le \ e^{\int_{t_0}^t a(r)\,dr} \lim_{\epsilon\to 0} g(\epsilon)(\frac{t_0}{\epsilon})^{k-\frac{1}{2}} = 0.$$

Thus, letting $\epsilon \to 0$ in (3.21) gives

$$g(t) = \int_0^t b(s)e^{\int_s^t a(r)\,dr}\,ds,\ 0 \le t < 1. \tag{3.22}$$

Using partial fractions, one finds that

$$\frac{(k-1)-(k-2)r}{r(1-r)} = \frac{k-1}{r} + \frac{1}{1-r}.$$

We also have

$$\frac{r^{k-1}}{(1-r)} = \frac{1}{1-r} - (1 + r + \cdots + r^{k-2}).$$

Thus, we can rewrite $a(r)$ from (3.20) as

$$a(r) = \frac{k-1}{r} + \frac{3}{1-r} - 2(1 + r + \cdots + r^{k-2}).$$

We then obtain

$$\int^t a(r)\,dr = (k-1)\log t - 3\log(1-t) - 2\sum_{j=1}^{k-1}\frac{t^j}{j},$$

and thus

$$e^{\int_s^t a(r)\,dr} = \big(t^{k-1}(1-t)^{-3}e^{-2\sum_{j=1}^{k-1}\frac{t^j}{j}}\big)\big(s^{1-k}(1-s)^3e^{2\sum_{j=1}^{k-1}\frac{s^j}{j}}\big). \tag{3.23}$$

Substituting this in (3.22) and recalling the definition of b from (3.20), we obtain

$$g(t) = \frac{t^{k-1}}{(1-t)^3}e^{-2\sum_{j=1}^{k-1}\frac{t^j}{j}}\int_0^t ke^{2\sum_{j=1}^{k-1}\frac{s^j}{j}}\,ds. \tag{3.24}$$

We see that g has a third-order singularity at $t = 1$. We proceed to "unzip" this singularity to reveal the answer to our problem.

We have the following proposition which connects the limiting behavior of H_n with that of S_n.

Proposition 3.1.

$$\lim_{n\to\infty} \frac{H_n^{(k)}}{n} = \ell$$

if and only if

$$\lim_{n\to\infty} \frac{S_n^{(k)}}{n^2} = \frac{\ell}{2}.$$

Proof. The proof is immediate from (3.11). □

And we have the following proposition which connects the limiting behavior of S_n with the singularity in g at $t = 1$.

Proposition 3.2. *If*

$$\lim_{n\to\infty} \frac{S_n^{(k)}}{n^2} = L,$$

then

$$\lim_{t\to 1}(1-t)^3 g(t) = 2L.$$

Proof. Since $\lim_{n\to\infty} \frac{S_n}{n^2} = L$, we also have $\lim_{n\to\infty} \frac{S_n}{n(n-1)} = L$. Let $\epsilon > 0$. Choose n_0 such that $|\frac{S_n}{n(n-1)} - L| \le \epsilon$, for $n > n_0$. Then recalling (3.15), we have

$$\sum_{n=0}^{n_0} S_n t^n + (L-\epsilon) \sum_{n=n_0+1}^{\infty} n(n-1)t^n \le g(t) \le \sum_{n=0}^{n_0} S_n t^n + (L+\epsilon) \sum_{n=n_0+1}^{\infty} n(n-1)t^n. \tag{3.25}$$

Now

$$\sum_{n=0}^{\infty} n(n-1)t^n = t^2 (\sum_{n=0}^{\infty} t^n)'' = t^2 (\frac{1}{1-t})'' = \frac{2t^2}{(1-t)^3},$$

so

$$\sum_{n=n_0+1}^{\infty} n(n-1)t^n = \frac{2t^2}{(1-t)^3} - \sum_{n=0}^{n_0} n(n-1)t^n.$$

Substituting this latter equality in (3.25), multiplying by $(1-t)^3$, and letting $t \to 1$, we obtain

$$2L - 2\epsilon \leq \liminf_{t\to 1}(1-t)^3 g(t) \leq \limsup_{t\to 1}(1-t)^3 g(t) \leq 2L + 2\epsilon.$$

As $\epsilon > 0$ is arbitrary, the proposition follows. □

In order to exploit Propositions 3.1 and 3.2, we will establish the existence of the limit $\lim_{n\to\infty} \frac{S_n}{n^2}$.

Proposition 3.3. $\lim_{n\to\infty} \frac{S_n^{(k)}}{n^2}$ *exists.*

Proof. Rewriting the recursion equation for S_n in (3.13) so that only S_n appears on the left hand side, then dividing both sides by n^2 and subtracting $\frac{S_{n-1}}{(n-1)^2}$ from both sides, we have

$$\frac{S_n}{n^2} - \frac{S_{n-1}}{(n-1)^2} = \frac{k}{n^2} + \frac{S_{n-1}}{n^2} - \frac{S_{n-1}}{(n-1)^2} + \frac{2S_{n-k}}{n^2(n-k+1)} =$$

$$\frac{k}{n^2} - \frac{2n-1}{n^2(n-1)^2}S_{n-1} + \frac{2S_{n-k}}{n^2(n-k+1)} =$$

$$\frac{k}{n^2} - \frac{2n-1}{n^2(n-1)^2}S_{n-1} + \frac{2S_{n-1}}{n^2(n-k+1)} - \frac{2}{n^2(n-k+1)}(H_{n-k+1} + \cdots + H_{n-1}) =$$

$$\frac{k}{n^2} + \frac{(2k-5)n+3-k}{n^2(n-1)^2(n-k+1)}S_{n-1} - \frac{2}{n^2(n-k+1)}(H_{n-k+1} + \cdots + H_{n-1}). \tag{3.26}$$

As already noted, from the definitions, we have $H_l \leq l$ and $S_l \leq \frac{1}{2}l(l+1)$. Thus, there exists a $C > 0$ such that

$$\left|\frac{(2k-5)n+3-k}{n^2(n-1)^2(n-k+1)}\right| S_{n-1} \leq \frac{C}{n^2} \text{ and}$$

$$\frac{2}{n^2(n-k+1)}(H_{n-k+1} + \cdots + H_{n-1}) \leq \frac{C}{n^2}. \tag{3.27}$$

This shows that the right hand side of (3.26) is $O(\frac{1}{n^2})$ and thus so is the left hand side. Consequently, the telescopic series $\sum_{n=2}^{\infty}\left(\frac{S_n}{n^2} - \frac{S_{n-1}}{(n-1)^2}\right)$ is convergent. Since

$$\frac{S_n}{n^2} = \sum_{j=2}^{n}\left(\frac{S_j}{j^2} - \frac{S_{j-1}}{(j-1)^2}\right),$$

we conclude that $\lim_{n\to\infty} \frac{S_n}{n^2}$ exists. □

By Propositions 3.1 and 3.3, $\ell := \lim_{n\to\infty} \frac{H_n}{n}$ exists. Then by Propositions 3.1 and 3.2 (with $L = \frac{\ell}{2}$), it follows that

$$\lim_{t\to 1}(1-t)^3 g(t) = \ell.$$

However, from the explicit formula for g in (3.24), we have

$$\lim_{t\to 1}(1-t)^3 g(t) = ke^{-2\sum_{j=1}^{k-1}\frac{1}{j}}\int_0^1 e^{2\sum_{j=1}^{k-1}\frac{s^j}{j}}\,ds = p_k.$$

Thus, $\ell = p_k$, completing the proof of (3.3).

We now turn to the proof of (3.4). We derive a formula by the method used to obtain (3.8). Recall the discussion preceding (3.8). Note that conditioned on B_j being chosen on the first step, the final state of the $j-1$ molecules to the left of B_j and the final state of the $n+1-j-k$ molecules to the right of B_j are independent of one another. Let $M_{j-1;k;1}$ and $M_{n+1-j-k;k;2}$ be independent random variables distributed according to the distributions of $M_{j-1;k}$ and $M_{n+1-j-k;k}$, respectively. Then similar to (3.8), we have

$$E(M_{n;k}^2|B_j \text{ selected first}) = E(k + M_{j-1;k;1} + M_{n+1-j-k;k;2})^2 =$$

$$k^2 + L_{j-1} + L_{n+1-j-k} + 2kH_{j-1} + 2kH_{n+1-j-k} + 2H_{j-1}H_{n+1-j-k}, \quad (3.28)$$

where the last term comes from the fact that the independence gives

$$EM_{j-1;k;1}M_{n+1-j-k;k;2} = EM_{j-1;k;1}EM_{n+1-j-k;k;2}.$$

Thus, similar to the passage from (3.8) to (3.9), we have

$$L_n = k^2 + \frac{2}{n-k+1}\sum_{j=0}^{n-k} L_j + \frac{4k}{n-k+1}\sum_{j=0}^{n-k} H_j + \frac{2}{n-k+1}\sum_{j=0}^{n-k} H_j H_{n-k-j},$$

$$\text{for } k \ge n. \quad (3.29)$$

We simplify the above recursion relation by defining

$$R_n = \sum_{j=0}^{n} L_j.$$

Of course, we have $L_n = 0$, for $n = 0, \ldots, k-1$, and thus,

$$R_n = 0, \ \text{for } n = 0, \ldots, k-1. \quad (3.30)$$

Recalling (3.10), we can now rewrite (3.29) in the form

$$R_n = R_{n-1} + k^2 + \frac{2}{n-k+1}R_{n-k} + \frac{4k}{n-k+1}S_{n-k} + \frac{2}{n-k+1}\sum_{j=0}^{n-k} H_j H_{n-k-j}, \ n \ge k. \quad (3.31)$$

Proposition 3.4.

$$\lim_{n\to\infty}\frac{1}{n^3}\sum_{j=0}^{n-k}H_j^{(k)}H_{n-k-j}^{(k)}=\frac{p_k^2}{6}.$$

Proof. Let $\epsilon > 0$. Since $\lim_{n\to\infty}\frac{H_n}{n}=p_k$, we can find an n_ϵ such that $(p_k-\epsilon)n\le H_n\le(p_k+\epsilon)n$, for $n>n_\epsilon$. Thus

$$\begin{aligned}&(p_k-\epsilon)^2\sum_{n_\epsilon<j<n-n_\epsilon-k}j(n-k-j)+\sum_{0\le j\le n_\epsilon,n-n_\epsilon-k\le j\le n-k}H_jH_{n-k-j}\le\\&\sum_{j=0}^{n-k}H_jH_{n-k-j}\le\\&(p_k+\epsilon)^2\sum_{n_\epsilon<j<n-n_\epsilon-k}j(n-k-j)+\sum_{0\le j\le n_\epsilon,n-n_\epsilon-k\le j\le n-k}H_jH_{n-k-j}.\end{aligned}\tag{3.32}$$

Since $H_j\le j$, for all j, we have

$$\sum_{0\le j\le n_\epsilon,n-n_\epsilon-k\le j\le n-k}H_jH_{n-k-j}\le 2(n_\epsilon+1)n_\epsilon n.\tag{3.33}$$

(There are $2(n_\epsilon+1)$ summands on the left hand side of (3.33), and each summand, H_jH_{n-k-j}, is less than or equal to $n_\epsilon n$.) Using the identity $\sum_{j=1}^n j^2=\frac{1}{6}n(n+1)(2n+1)$, we have

$$\begin{aligned}&\sum_{1\le j<n-n_\epsilon-k}j(n-k-j)=(n-k)\sum_{1\le j<n-n_\epsilon-k}j-\sum_{1\le j<n-n_\epsilon-k}j^2=\\&\frac{1}{2}(n-k)(n-n_\epsilon-k-1)(n-n_\epsilon-k)-\\&\frac{1}{6}(n-n_\epsilon-k-1)(n-n_\epsilon-k)\big(2(n-n_\epsilon-k-1)+1\big)=\frac{1}{6}n^3+o(n^3),\ \text{as } n\to\infty.\end{aligned}\tag{3.34}$$

Of course,

$$\sum_{1\le j\le n_\epsilon}j(n-k-j)\le n\sum_{1\le j\le n_\epsilon}j\le\frac{1}{2}nn_\epsilon(n_\epsilon+1).\tag{3.35}$$

From (3.32)–(3.35), we conclude that

$$\frac{1}{6}(p_k-\epsilon)^2\le\liminf_{n\to\infty}\frac{1}{n^3}\sum_{j=0}^{n-k}H_jH_{n-k-j}\le\limsup_{n\to\infty}\frac{1}{n^3}\sum_{j=0}^{n-k}H_jH_{n-k-j}\le\frac{1}{6}(p_k+\epsilon)^2,$$

which completes the proof, since $\epsilon>0$ is arbitrary. □

We can rewrite (3.31) as

$$R_n = \frac{n-k+3}{n-k+1}R_{n-1} + k^2 - \frac{2}{n-k+1}(L_{n-k+1} + \cdots + L_{n-1}) +$$

$$\frac{4k}{n-k+1}S_{n-k} + \frac{2}{n-k+1}\sum_{j=0}^{n-k} H_j H_{n-k-j}.$$

Since $L_j \le j^2$ and $S_j \le \frac{1}{2}j(j+1)$, we conclude from Proposition 3.4 that R_n satisfies an equation of the form

$$R_n = \frac{n-k+3}{n-k+1}R_{n-1} + W_n, \text{ where } W_n \text{ satisfies } \lim_{n\to\infty} \frac{W_n}{n^2} = \frac{p_k^2}{3}. \tag{3.36}$$

In Exercise 3.1 the reader is asked to show that if for some n_0, the positive sequence $\{\hat{R}_n\}_{n=n_0}^{\infty}$ satisfies $\hat{R}_n \le \frac{n-k+3}{n-k+1}\hat{R}_{n-1} + cn^2$ ($\hat{R}_n \ge \frac{n-k+3}{n-k+1}\hat{R}_{n-1} + cn^2$), then $\limsup_{n\to\infty} \frac{\hat{R}_n}{n^3} \le c$ ($\liminf_{n\to\infty} \frac{\hat{R}_n}{n^3} \ge c$). Using this with (3.36), we conclude that

$$\lim_{n\to\infty} \frac{R_n}{n^3} = \frac{p_k^2}{3}. \tag{3.37}$$

Writing (3.31) in the form

$$L_n = k^2 + \frac{2}{n-k+1}R_{n-k} + \frac{4k}{n-k+1}S_{n-k} + \frac{2}{n-k+1}\sum_{j=0}^{n-k} H_j H_{n-k-j},\ n \ge k, \tag{3.38}$$

dividing both sides of this equation by n^2, and using (3.37), Proposition 3.4, and the fact that S_n is on the order n^2, we conclude that

$$\lim_{n\to\infty} \frac{L_n}{n^2} = 2\frac{p_k^2}{3} + 2\frac{p_k^2}{6} = p_k^2.$$

This gives (3.4) and completes the proof of Theorem 3.1. □

Exercise 3.1. Show that if for some n_0, the positive sequence $\{\hat{R}_n\}_{n=n_0}^{\infty}$ satisfies $\hat{R}_n \le \frac{n-k+3}{n-k+1}\hat{R}_{n-1} + cn^2$ ($\hat{R}_n \ge \frac{n-k+3}{n-k+1}\hat{R}_{n-1} + cn^2$), then $\limsup_{n\to\infty} \frac{\hat{R}_n}{n^3} \le c$ ($\liminf_{n\to\infty} \frac{\hat{R}_n}{n^3} \ge c$).

Exercise 3.2. Any molecule that remains unbonded at the end of the nearest neighbor k-tuple bonding process occurs in a maximal row of j unbonded molecules, for some $j \in [k-1]$. In the limit as $n \to \infty$, what fraction of molecules ends up in a maximal row of j unbounded molecules? Let's denote these fractions by $q_{k;j}$, $j \in [k-1]$. Of course $\sum_{j=1}^{k-1} q_{k;j} = 1 - p_k$.

(a) Let $k \geq 3$ and fix $j \in [k-1]$. Consider the following bonding process: implement the bonding of nearest neighbor k-tuples as described in the chapter. When this process terminates, bond all the unbonded molecules that occur in a maximal row of j unbonded molecules, but leave untouched all unbonded molecules that occur in a maximal row of i unbonded molecules, for some $i \neq j$. Let $M_{n;k,j}$ denote the number of bonded molecules at the end of the process, and let $H_n^{(k,j)} = EM_{n;k,j}$. Let $S_n^{(k,j)} = \sum_{i=0}^n H_i^{(k,j)}$. Convince yourself that $\{H_n^{(k,j)}\}_{n=0}^\infty$ satisfies the recursion equation (3.9) and that it satisfies the boundary condition (3.7) with one change, namely $H_j^{(k,j)} = j$, instead of $H_j^{(k,j)} = 0$. Thus, $\{S_n^{(k,j)}\}_{n=0}^\infty$ satisfies the recursion equation (3.13), and in place of the boundary condition (3.12), it satisfies the boundary condition $S_n^{(k,j)} = 0, n = 0, \ldots, j-1$; $S_n^{(k,j)} = j, n = j, \ldots, k-1$.

(b) Let $g_j(t) = \sum_{n=0}^\infty S_n^{(k,j)} t^n$ denote the generating function for $\{S_n^{(k,j)}\}_{n=0}^\infty$. Show that g_j solves the differential equation $g_j'(t) = a(t)g_j(t) + b_j(t)$, where a is as in (3.20) and

$$b_j(t) = b(t) + \frac{-j(k-1-j)t^{j-1} + j(k-j)t^j - jt^{k-1}}{(1-t)^3},$$

with b as in (3.20).

(c) In particular, note that $b_{k-1} = b$; therefore, g_{k-1} satisfies the same differential equation satisfied by g. Thus, (3.21) holds for g_{k-1}; that is,

$$g_{k-1}(t) = g_{k-1}(\epsilon)e^{\int_\epsilon^t a(r)\,dr} + \int_\epsilon^t b(s)e^{\int_s^t a(r)\,dr}\,ds, \; t \in (\epsilon, 1).$$

Use the fact that $g_{k-1}(\epsilon) = (k-1)\epsilon^{k-1} + O(\epsilon^k)$, as $\epsilon \to 0$, along with (3.23) to show that

$$\lim_{\epsilon \to 0} g_{k-1}(\epsilon)e^{\int_\epsilon^t a(r)\,dr} = (k-1)\frac{t^{k-1}}{(1-t)^3}e^{-2(t+\frac{t^2}{2}+\cdots+\frac{t^{k-1}}{k-1})}.$$

(d) Use (c) to show that

$$q_{k;k-1} = (k-1)e^{-2(1+\frac{1}{2}+\cdots+\frac{1}{k-1})}.$$

In particular then, $q_{3,2} = 2e^{-3} \approx 0.0996 \approx 0.100$, and consequently $q_{3,1} = 1 - p_3 - q_{3,2} \approx 1 - 0.8237 - 0.0996 \approx 0.077$.

(e) It is well known that $\lim_{n\to\infty}\left(\sum_{r=1}^n \frac{1}{r} - \log n\right)$ exists; the limit is called *Euler's constant* and is denoted by γ. One has $\gamma \approx 0.5772$. For a proof, see, for example, [25]. Show that

$$q_{k;k-1} \sim \frac{1}{k-1}e^{-2\gamma}, \text{ as } k \to \infty.$$

(f) For $j \in [k-2]$, one obtains

$$g_j(t) = g_j(\epsilon)e^{\int_\epsilon^t a(r)\,dr} + \int_\epsilon^t b_j(s)e^{\int_s^t a(r)\,dr}\,ds,\ t \in (\epsilon, 1). \tag{3.39}$$

Show that since $g_j(\epsilon) = j\epsilon^j + O(\epsilon^{j+1})$, as $\epsilon \to 0$, one has

$$\lim_{\epsilon \to 0} g_j(\epsilon)e^{\int_\epsilon^t a(r)\,dr} = \infty.$$

On the other hand, since b_j appears instead of $b_{k-1} = b$, show that one also has

$$\lim_{\epsilon \to 0} \int_\epsilon^t b_j(s)e^{\int_s^t a(r)\,dr}\,ds = -\infty.$$

You are invited to show that the appropriate terms in $g_j(\epsilon)e^{\int_\epsilon^t a(r)\,dr}$ and $\int_\epsilon^t b_j(s)e^{\int_s^t a(r)\,dr}\,ds$ cancel each other out and to obtain a finite limiting expression as $\epsilon \to 0$ on the right hand side of (3.39). This limiting expression is then also $g_j(t)$. One then has $\lim_{t \to 1}(1-t)^3 g_j(t) = p_k + q_{k;j}$, which gives an explicit formula for $q_{k;j}$. The above analysis gets more involved the smaller j is. Try it first for $j = k-2$.

Chapter Notes

The calculation of (3.1) in the case $k = 2$ goes back to an article by the Nobel Prize winning chemist Flory in 1939 [21]. The problem was rediscovered by Page, who obtained the asymptotic behavior for the mean and variance in the case $k = 2$ [28]. The method used there does not generalize to $k > 2$. Theorem 3.1 seems to be new. A continuous space version of this problem was considered by Rényi [31].

Chapter 4
The Arcsine Laws for the One-Dimensional Simple Symmetric Random Walk

The *simple, symmetric random walk* $\{S_n\}_{n=0}^{\infty}$ on $\mathbb{Z}$ starts at step $n = 0$ at $0 \in \mathbb{Z}$ and at each successive step jumps one unit to the right or left, each with probability $\frac{1}{2}$. The random walk is called "simple" because the sizes of its jumps are restricted to the set $\{1, -1\}$. One way to realize this random walk is as follows. Let $\{X_n\}_{n=1}^{\infty}$ be an infinite sequence of independent, identically distributed random variables distributed according to the Bernoulli distribution with parameter $\frac{1}{2}$; that is, $P(X_j = 1) = P(X_j = -1) = \frac{1}{2}$. Now define $S_0 = 0$ and $S_n = \sum_{j=1}^{n} X_j$, $n \geq 1$.

We begin with a fundamental fact about the simple, symmetric random walk on $\mathbb{Z}$.

Proposition 4.1.

$$P(\limsup_{n\to\infty} S_n = \infty \text{ and } \liminf_{n\to\infty} S_n = -\infty) = 1. \tag{4.1}$$

Remark 1. A moment's thought shows that (4.1) is equivalent to the statement that the random walk is *recurrent*; that is, with probability one, $\{S_n\}_{n=0}^{\infty}$ visits every site in $\mathbb{Z}$ infinitely often.

Remark 2. One can consider a simple, symmetric random walk $\{S_n\}_{n=0}^{\infty}$ on $\mathbb{Z}^d$, the d-dimensional lattice—at each step it jumps in one of the $2d$ directions with probability $\frac{1}{2d}$. Again, the random walk is called recurrent if with probability one every site is visited infinitely often. It is called *transient* if $P(\lim_{n\to\infty} |S_n| = \infty) = 1$. In 1923, G. Polya proved the quite surprising result that this random walk is recurrent in two dimensions but transient in three or more dimensions. For a proof of this, see, for example, [15].

Proof. By Remark 1 above, to prove the proposition, it suffices to prove that with probability one, the random walk visits every site in $\mathbb{Z}$ infinitely often. Let p denote the probability that the random walk $\{S_n\}_{n=0}^{\infty}$ ever returns to its starting point 0. We will show that $p = 1$. Let N_0 denote the number of times the random walk is at 0 after time $n = 0$. Then of course, $P(N_0 = 0) = 1 - p$. Now let's calculate

R.G. Pinsky, *Problems from the Discrete to the Continuous*, Universitext,
DOI 10.1007/978-3-319-07965-3_4, © Springer International Publishing Switzerland 2014

$P(N_0 = 1)$. In order to have $N_0 = 1$, the random walk must return to 0 and then never return to 0 again. The probability of returning to 0 is p. If the random walk returns to 0, it continues independently of everything that has already transpired. Thus, conditioned on returning to 0, the probability that the random walk does not return to 0 again is $1 - p$. So $P(N_0 = 1) = p(1 - p)$. Continuing with this line of reasoning, we obtain

$$P(N_0 = n) = p^n(1 - p),\ n = 0, 1, \ldots.$$

If $p = 1$, it follows from the above reasoning that $P(N_0 = \infty) = 1$; that is, with probability one, the random walk visits 0 infinitely often. If $p \in (0, 1)$, then the above calculation shows that N_0 is distributed according to the *geometric distribution* with parameter p. For $p \in (0, 1)$, the expected value EN_0 of N_0 is given by

$$EN_0 = \sum_{n=0}^{\infty} nP(N_0 = n) = \sum_{n=0}^{\infty} np^n(1 - p) = p(1 - p)\sum_{n=0}^{\infty} np^{n-1} =$$
$$p(1 - p)\frac{d}{dp}\Big(\sum_{n=0}^{\infty} p^n\Big) = p(1 - p)\Big(\frac{1}{1 - p}\Big)' = \frac{p}{1 - p}. \tag{4.2}$$

(The term by term differentiation above is permitted because for any $p_0 < 1$, the series is uniformly absolutely convergent over $p \in [0, p_0]$.) Of course, if $p = 1$, then $EN_0 = \infty$. Thus, the formula for EN_0 in (4.2) also holds if $p = 1$.

We now calculate EN_0 in a different way. Let $1_{\{S_n=0\}}$ denote the *indicator random variable* that is equal to 1 if $S_n = 0$ and is equal to 0 otherwise. Then N_0, the number of times the random walk returns to 0, can be represented as

$$N_0 = \sum_{n=1}^{\infty} 1_{\{S_n=0\}}.$$

By the linearity of the expectation and the nonnegativity of the summands, we conclude that

$$EN_0 = \sum_{n=1}^{\infty} P(S_n = 0), \tag{4.3}$$

since $E1_{\{S_n=0\}} = 0 \cdot P(S_n \neq 0) + 1 \cdot P(S_n = 0) = P(S_n = 0)$.

Since the random walk starts at 0, it can only return to 0 at even times; thus, $P(S_{2n+1} = 0) = 0$. Since the random walk has two equally likely choices at each step, there are 2^{2n} equally likely paths that the random walk can traverse during its first $2n$ steps. Now one has $S_{2n} = 0$ if and only if from among the first $2n$ jumps, n of them were to the right and n of them were to the left. There are $\binom{2n}{n}$ such paths;

thus,

$$P(S_{2n}=0)=\frac{\binom{2n}{n}}{2^{2n}}. \tag{4.4}$$

Using Stirling's formula, namely, $n! \sim n^n e^{-n}\sqrt{2\pi n}$ as $n \to \infty$, we have

$$\frac{\binom{2n}{n}}{2^{2n}}=\frac{(2n)!}{(n!)^2 2^{2n}}\sim\frac{(2n)^{2n}e^{-2n}\sqrt{4\pi n}}{n^{2n}e^{-2n}(2\pi n)2^{2n}}=\frac{1}{\sqrt{\pi n}},\ \text{as } n\to\infty. \tag{4.5}$$

Since $\sum_{n=1}^{\infty}\frac{1}{\sqrt{n}}=\infty$, it follows from (4.3)–(4.5) that $EN_0=\infty$. In light of (4.2), we conclude that $p=1$.

We have shown that with probability one, the random walk returns to 0. Upon returning to 0, the random walk continues independently of everything that transpired previously; thus, in fact, with probability one, the random walk visits 0 infinitely often. From this, it is easy to show that in fact with probability one the random walk visits every site infinitely often. We leave this as Exercise 4.1. □

Define

$$T_0=\inf\{n>0: S_n=0\}.$$

The random time T_0 is called the *first return time to 0*. By Proposition 4.1, it follows that $P(T_0<\infty)=1$. However, perhaps surprisingly, one has $ET_0=\infty$; the reader is guided through a proof of this in Exercise 4.2. This result suggests that there is quite some tendency for the random walk to take a long time to return to 0. In this chapter we present two results which give vivid expression to this phenomenon.

The *arcsine distribution* will figure prominently in the results of this chapter. The distribution function for this distribution is defined by

$$F_{\text{arcsin}}(x)=\frac{2}{\pi}\arcsin\sqrt{x},\ \ 0\le x\le 1.$$

The corresponding density function $f_{\text{arcsin}}(x)=F'_{\text{arcsin}}(x)$ is given by

$$f_{\text{arcsin}}(x)=\frac{1}{\pi}\frac{1}{\sqrt{x(1-x)}},\ \ 0<x<1.$$

Our first theorem concerns the random time

$$L_0^{(2n)}=\max\{k\le 2n: S_k=0\},$$

which is the *last return time* to 0 up to step $2n$. By parity considerations, $L_0^{(2n)}$ can take on only even values.

Theorem 4.1.

$$P(L_0^{(2n)} = 2k) = \frac{\binom{2k}{k}\binom{2n-2k}{n-k}}{2^{2n}}, \quad k = \{0, 1, \ldots, n\}. \tag{4.6}$$

Furthermore,

$$\lim_{n\to\infty} P(\frac{L_0^{(2n)}}{2n} \le x) = \frac{2}{\pi} \arcsin\sqrt{x}, \quad 0 \le x \le 1. \tag{4.7}$$

Remark. This theorem highlights the tendency of the random walk to take a long time to return to 0. Indeed, since the density $f_{\text{arcsin}}(x)$ blows up at $x = 0, 1$, it follows from (4.7) that for large n the most likely epochs k for the last visit to 0 up to time $2n$ are those satisfying $k = o(n)$ or $k = 2n - o(n)$, that is, those epochs at the very beginning or at the very end of the trajectory. Since $\frac{2}{\pi} \arcsin\sqrt{\frac{1}{2}} = \frac{1}{2}$, from (4.7) it also follows that for large n, there is a probability of about $\frac{1}{2}$ that a random walk trajectory of $2n$ steps will never return to 0 during the second half of its life.

Our second theorem concerns the random variable O_{2n}^+, which should be thought of as the number of steps $k \in [2n]$ at which the random walk is positive (or nonnegative). Of course, the number of steps between 1 and $2n$ that the random walk is positive is usually not equal to the number of steps that it is nonnegative. In order to obtain an exact result in closed form for all n, we need to work in a symmetric setting. Therefore, if the random walk is equal to 0 at some step $2k$, we classify that step as "positive" if the previous step was positive and "negative" if the previous step was negative. That is,

$$O_{2n}^+ = |\{k \in [2n] : S_k > 0 \text{ or } S_k = 0 \text{ and } S_{k-1} > 0\}|.$$

We call O_{2n}^+ the *occupation time of the positive half line* up to time $2n$. Then $\frac{O_{2n}^+}{2n}$ gives the fraction of steps among the first $2n$ steps that the random walk is in the positive half line. Note that by parity considerations, O_{2n}^+ can only take on even values.

Theorem 4.2.

$$P(O_{2n}^+ = 2k) = \frac{\binom{2k}{k}\binom{2n-2k}{n-k}}{2^{2n}}, \quad k = \{0, 1, \ldots, n\}. \tag{4.8}$$

Furthermore,

$$\lim_{n\to\infty} P(\frac{O_{2n}^+}{2n} \le x) = \frac{2}{\pi} \arcsin\sqrt{x}, \quad 0 \le x \le 1. \tag{4.9}$$

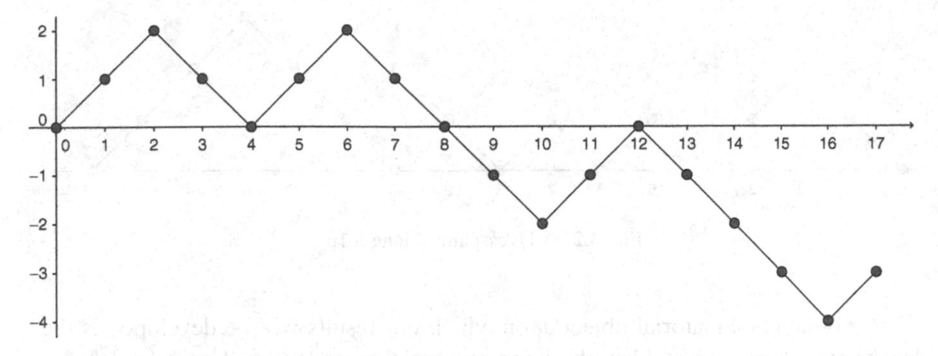

Fig. 4.1 A random walk path of length 17

Remark 1. Since the density $f_{\text{arcsin}}(x)$ takes on its minimum at $x = \frac{1}{2}$, and since it blows up at $x = 0, 1$, it follows that for large n the most likely percentages of time that a random walk trajectory is nonnegative are around 0 % and 100 %, while the least likely percentage is around 50 %! To put it in a different way, if two players bet a dollar each on a succession of fair coin flips, then after a long time it is overwhelmingly more likely that one of the players was leading almost the whole time than that each player was leading about half the time. This result even more vividly highlights the tendency of the random walk to take a long time to return to 0.

Remark 2. Let $O_{2n}^0 = \{k \in [n] : S_{2k} = 0\}$ denote the number of visits to 0 of the random walk up to step $[2n]$. It is not hard to show that the random variable $\frac{O_{2n}^0}{2n}$, denoting the fraction of steps up to $2n$ at which the random walk is at 0, converges to 0 in probability; that is,

$$\lim_{n\to\infty} P(\frac{O_{2n}^0}{2n} > \epsilon) = 0, \text{ for all } \epsilon > 0. \tag{4.10}$$

We leave this as Exercise 4.3. In light of this, it follows that (4.9) would also hold if we had defined O_{2n}^+ in an asymmetric fashion as the number of steps up to $[2n]$ for which the random walk is nonnegative: $|\{k \in [2n] : S_k \geq 0\}|$.

Our approach to proving the above two theorems will be completely combinatorial rather than probabilistic. Generating functions will play a seminal role. A *random walk path* of length m is a path $\{x_j\}_{j=0}^m$ which satisfies

$$\begin{aligned} &x_0 = 0; \\ &x_j - x_{j-1} = \pm 1, \ j \in [m]; \end{aligned} \tag{4.11}$$

See Fig. 4.1. Since a random walk path has two choices at each step, there are 2^m random walk paths of length m. The probability that the simple, symmetric random walk behaves in a certain way up until time m is simply the number of random walk paths that behave in that certain way divided by 2^m.

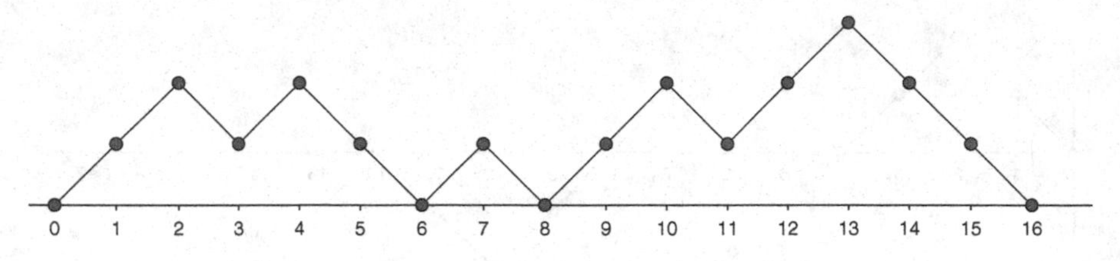

Fig. 4.2 A Dyck path of length 16

Our basic combinatorial object upon which our results will be developed is the *Dyck path*. A Dyck path of length $2n$ is a nonnegative random walk path $\{x_j\}_{j=0}^{2n}$ of length $2n$ which returns to 0 at step $2n$; that is, in addition to satisfying (4.11) with $m = 2n$, it also satisfies the following conditions:

$$\begin{aligned} &x_j \geq 0, \ j \in [2n]; \\ &x_{2n} = 0. \end{aligned} \tag{4.12}$$

See Fig. 4.2. We use generating functions to determine the number of Dyck paths. Let d_n denote the number of Dyck paths of length $2n$. We also define $d_0 = 1$.

Proposition 4.2. *The number of Dyck paths of length* $2n$ *is given by*

$$d_n = \frac{1}{n+1}\binom{2n}{n}, \ n \geq 1.$$

Remark. The number $C_n := \frac{1}{n+1}\binom{2n}{n}$ is known as the nth *Catalan* number.

Proof. We derive a recursion formula for $\{d_n\}_{n=0}^{\infty}$. A *primitive Dyck path* of length $2k$ is a Dyck path $\{x_j\}_{j=0}^{2k}$ of length $2k$ which satisfies $x_j > 0$ for $j = 1, \ldots, 2k-1$. Let v_k denote the number of primitive Dyck paths of length $2k$. Every Dyck path of length $2n$ returns to 0 for the first time at $2k$, for some $k \in [n]$. Consider a Dyck path of length $2n$ that returns to 0 for the first time at $2k$. The part of the path from time 0 to time $2k$ is a primitive Dyck path of length $2k$, and the part of the path from time $2k$ to $2n$ is an arbitrary Dyck path of length $2n - 2k$. (In Fig. 4.2, the Dyck path of length 16 is composed of an initial primitive Dyck path of length 6, followed by a Dyck path of length 10.) This reasoning yields the recurrence relation

$$d_n = \sum_{k=1}^{n} v_k d_{n-k}, \ n \geq 1. \tag{4.13}$$

Now we claim that

$$v_k = d_{k-1}, \ k \geq 1. \tag{4.14}$$

Indeed, a primitive Dyck path $\{x_j\}_{j=0}^{2k}$ must satisfy $x_1 = 1, x_j \geq 1$, for $j \in [2k-2]$, $x_{2k-1} = 1$, $x_{2k} = 0$. Thus, letting $y_j = x_{j+1} - 1$, $0 \leq j \leq 2k-2$, it follows that $\{y_j\}_{j=0}^{2k-2}$ is a Dyck path. Of course, this analysis can be reversed. This shows that there is a 1-1 correspondence between primitive Dyck paths of length $2k$ and arbitrary Dyck paths of length $2(k-1)$, proving (4.14). From (4.13) and (4.14) we obtain the Dyck path recursion formula

$$d_n = \sum_{k=1}^{n} d_{k-1} d_{n-k}. \tag{4.15}$$

Let

$$D(x) = \sum_{n=0}^{\infty} d_n x^n \tag{4.16}$$

be the generating function for $\{d_n\}_{n=0}^{\infty}$. Since there are 2^{2n} random walk paths of length $2n$, we have the trivial estimate $d_n \leq 2^{2n} = 4^n$. Thus, the power series defining $D(x)$ is absolutely convergent for $|x| < \frac{1}{4}$. The product of two absolutely convergent power series $\sum_{n=0}^{\infty} a_n x^n$ and $\sum_{n=0}^{\infty} b_n x^n$ is $\sum_{n=0}^{\infty} c_n x^n$, where $c_n = \sum_{j=0}^{n} a_j b_{n-j}$. Thus, if in (4.15), the term d_{k-1} were d_k instead, and the summation started from $k = 0$ instead of from $k = 1$, then we would have had $D^2(x) = D(x)$. As it is, we "correct" for these deficiencies by multiplying by x and adding 1: it is easy to check that (4.16) and (4.15) give

$$D(x) = xD^2(x) + 1. \tag{4.17}$$

Solving this quadratic equation in D gives $D(x) = \frac{1 \pm \sqrt{1-4x}}{2x}$. Since we know from (4.16) that $D(0) = 1$, we conclude that the generating function for $\{d_n\}_{n=0}^{\infty}$ is given by

$$D(x) = \frac{1 - \sqrt{1-4x}}{2x}, \ |x| < \frac{1}{4}. \tag{4.18}$$

Now $(1-4x)^{\frac{1}{2}}|_{x=0} = 1$, $((1-4x)^{\frac{1}{2}})'|_{x=0} = -2$ and

$$\frac{1}{n!}((1-4x)^{\frac{1}{2}})^{(n)}|_{x=0} = -\frac{1}{n!} 2^n \prod_{j=1}^{n-1} (2j-1) = -\frac{2^n (2n-2)!}{n! 2^{n-1} (n-1)!} =$$

$$-\frac{2}{2n-1}\binom{2n-1}{n}, \text{ for } n \geq 2;$$

thus, the Taylor series for $\sqrt{1-4x}$ is given by

$$\sqrt{1-4x} = 1 - 2x - \sum_{n=2}^{\infty} \frac{2}{2n-1}\binom{2n-1}{n} x^n. \tag{4.19}$$

The coefficient of x^{n+1} in (4.19) is $-\frac{2}{2n+1}\binom{2n+1}{n+1} = -\frac{2}{n+1}\binom{2n}{n}$. Using this along with (4.18) and (4.19), we conclude that

$$D(x) = \sum_{n=0}^{\infty} \frac{1}{n+1}\binom{2n}{n} x^n, \ |x| < \frac{1}{4}. \tag{4.20}$$

From (4.20) and (4.16) it follows that $d_n = \frac{1}{n+1}\binom{2n}{n}$. □

The proof of the proposition gives us the following corollary.

Corollary 4.1. *The generating function for the sequence* $\{d_n\}_{n=0}^{\infty}$, *which counts Dyck paths, is given by*

$$D(x) = \frac{1-\sqrt{1-4x}}{2x}, \ |x| < \frac{1}{4}.$$

Let w_n denote the number of nonnegative random walk paths of length $2n$. The difference between such a path and a Dyck path is that for such a path there is no requirement that it return to 0 at time $2n$. We also define $w_0 = 1$. We now calculate $\{w_n\}_{n=0}^{\infty}$ by deriving a recursion formula which involves $\{d_n\}_{n=0}^{\infty}$.

Proposition 4.3. *The number* w_n *of nonnegative random walk paths of length* $2n$ *is given by*

$$w_n = \binom{2n}{n}, \ n \geq 1. \tag{4.21}$$

Remark. The number of random walk paths of length $2n$ that return to 0 at time $2n$ is also given by $\binom{2n}{n}$, since to obtain such a path, we must choose n jumps of $+1$ and n jumps of -1. Thus, we have the following somewhat surprising corollary.

Corollary 4.2.

$$P(S_1 \geq 0, \ldots, S_{2n} \geq 0) = P(S_{2n} = 0).$$

Proof of Proposition 4.3. Of course every nonnegative random walk path of length $2n+2$, when restricted to its first $2n$ steps, constitutes a nonnegative random walk path of length $2n$. A nonnegative random walk path of length $2n$ which does not return to 0 at time $2n$, that is, which is not a Dyck path, can be extended in four different ways to create a nonnegative random walk path of length $2n+2$. On the other hand, a nonnegative random walk path of length $2n$ which is a Dyck path can only be extended in two different ways to create a nonnegative random walk path of length $2n + 2$. Thus, we have the relation

$$w_{n+1} = 4(w_n - d_n) + 2d_n = 4w_n - 2d_n, \ n \geq 0. \tag{4.22}$$

Let

$$W(x) = \sum_{n=0}^{\infty} w_n x^n$$

be the generating function for $\{w_n\}_{n=0}^{\infty}$. As with the power series defining $D(x)$, it is clear that the power series defining $W(x)$ converges for $|x| < \frac{1}{4}$. Multiply equation (4.22) by x^n and sum over n from 0 to ∞. On the left side we obtain $\sum_{n=0}^{\infty} w_{n+1}x^n = \frac{1}{x}(W(x)-1)$, and on the right hand side we obtain $4W(x)-2D(x)$. From the resulting equation, $\frac{1}{x}(W(x) - 1) = 4W(x) - 2D(x)$, we obtain

$$W(x) = \frac{1 - 2xD(x)}{1 - 4x}. \tag{4.23}$$

Substituting for $D(x)$ in (4.23) from Corollary 4.1, we obtain

$$W(x) = \frac{1}{\sqrt{1 - 4x}}, \ |x| < \frac{1}{4}.$$

We have $W(0) = 1$, and for $n \geq 1$,

$$\frac{W^{(n)}(0)}{n!} = \frac{1}{n!}(1 - 4x)^{-\frac{1}{2}})^{(n)}|_{x=0} = \frac{1}{n!}2^n \prod_{j=1}^{n}(2j - 1) = \frac{1}{n!}2^n \frac{(2n)!}{2^n n!} = \binom{2n}{n}.$$

Thus the Taylor series for $W(x)$ is given by

$$W(x) = \sum_{n=0}^{\infty} \binom{2n}{n} x^n,$$

and we conclude that $w_n = \binom{2n}{n}$. □

Armed with Propositions 4.2 and 4.3, we can give a quick proof of (4.6).

Proof of Theorem 4.1. By the remark after Proposition 4.3, it follows that (4.6) holds for $k = n$. So we now assume that $k \in \{0, 1, \ldots, n - 1\}$. Given a random walk path, $\{x_j\}_{j=0}^{l}$, we define the negative of the path to be the path $\{-x_j\}_{j=0}^{l}$. If a random walk path of length $2n$ satisfies $L_0^{(2n)} = 2k$, then its first $2k$ steps constitute a random walk path that returns to 0 at time $2k$, and its last $2n - 2k$

steps constitute either a random walk path that is strictly positive or the negative of such a path. As noted in the remark after Proposition 4.3, there are $\binom{2k}{k}$ random walk paths of length $2k$ that return to 0 at time $2k$. How many strictly positive random walk paths of length $2n-2k$ are there? Let $\{x_j\}_{j=0}^{2n-2k}$ be such a path. Then $x_1 = 1$, and by parity considerations, $x_{2n-2k} \geq 2$. Consider now the part of the path from time 1 to time $2n-2k$. If we relabel and subtract one, $y_j = x_{j+1} - 1$, $j = 0, 1 \ldots, 2n-2k-1$, then we obtain a nonnegative random walk path of length $2n-2k-1$. By defining $y_{2n-2k} = y_{2n-2k-1} \pm 1$, we can extend this path in two ways to get a nonnegative random walk path of length $2n-2k$. This reasoning shows that there is a two-to-one correspondence between nonnegative random walk paths of length $2n-2k$ and strictly positive random walk paths of length $2n-2k$. We know that there are $w_{n-k} = \binom{2n-2k}{n-k}$ nonnegative random walk paths of length $2n-2k$; thus, we conclude that the number of strictly positive random walk paths of length $2n-2k$ is equal to $\frac{1}{2}\binom{2n-2k}{n-k}$. We conclude from the above analysis that the number of random walk paths of length $2n$ that satisfy $L_0^{(2n)} = 2k$ is equal to $\binom{2k}{k}\binom{2n-2k}{n-k}$, from which (4.6) follows.

We now consider (4.7). In Exercise 4.4 the reader is asked to apply Stirling's formula and show that for any $\epsilon > 0$,

$$\frac{\binom{2k}{k}\binom{2n-2k}{n-k}}{2^{2n}} \sim \frac{1}{\pi}\frac{1}{\sqrt{k(n-k)}}, \text{ uniformly over } \epsilon n \leq k \leq (1-\epsilon)n, \text{ as } n \to \infty. \tag{4.24}$$

Using (4.24) and (4.6), we have for $0 < a < b < 1$

$$P(a < \frac{L_0^{(2n)}}{2n} \leq b) = \sum_{k=[na]+1}^{[nb]} \frac{\binom{2k}{k}\binom{2n-2k}{n-k}}{2^{2n}} \sim \sum_{k=[na]+1}^{[nb]} \frac{1}{\pi}\frac{1}{\sqrt{k(n-k)}} =$$
$$\frac{1}{\pi}\sum_{k=[na]+1}^{[nb]} \frac{1}{\sqrt{\frac{k}{n}(1-\frac{k}{n})}}\frac{1}{n}, \text{ as } n \to \infty. \tag{4.25}$$

But the last term on the right hand side of (4.25) is a Riemann sum for $\frac{1}{\pi}\int_a^b \frac{1}{\sqrt{x(1-x)}}\,dx$. Thus, letting $n \to \infty$ in (4.25) gives

$$\lim_{n\to\infty} P(a < \frac{L_0^{(2n)}}{2n} \leq b) = \frac{1}{\pi}\int_a^b \frac{1}{\sqrt{x(1-x)}}\,dx = \frac{2}{\pi}\arcsin\sqrt{b} - \frac{2}{\pi}\arcsin\sqrt{a},$$

for $0 < a < b < 1$,

which is equivalent to (4.7). This completes the proof of Theorem 4.1. □

We now turn to the proof of Theorem 4.2.

Proof of Theorem 4.2. We need to prove (4.8). Of course, (4.9) follows from (4.8) just like (4.7) followed from (4.6). Recalling the symmetric definition of O_{2n}^+, for the purpose of this proof, we will refer to S_{2k} as "positive" if either $S_{2k} > 0$ or $S_{2k} = 0$ and $S_{2k-1} > 0$. Let $c_{n,k}$ denote the number of random walk paths of length $2n$ which are positive at exactly $2k$ steps. Since there are 2^{2n} random walk paths of length $2n$, in order to prove (4.8), we need to prove that

$$c_{n,k} = \binom{2k}{k}\binom{2n-2k}{n-k}, \; k = 0, 1, \ldots, n. \tag{4.26}$$

By Proposition 4.3, we have $c_{n,n} = \binom{2n}{n}$, and by symmetry, $c_{n,0} = \binom{2n}{n}$; thus, (4.26) holds for $k = 0, n$.

Consider now $k \in [n-1]$. A random walk path that satisfies $O_{2n}^+ = 2k$ must return to 0 before step $2n$. Consider the first return to 0. If the path was positive before the first return to 0, then the first return to 0 must occur at step $2j$, for some $j \in [k]$ (for otherwise, the path would be positive for more than $2k$ steps). If the path was negative before the first return to 0, then the first return to 0 must occur at step $2j$, for some $j \in [n-k]$ (for otherwise the path would be positive for fewer than $2k$ steps). In light of these facts, and recalling that $v_j = d_{j-1}$ is the number of primitive Dyck paths of length $2j$, it follows that for $j \in [k]$, the number of random walk paths of length $2n$ which start out positive, return to 0 for the first time at step $2j$, and are positive for exactly $2k$ steps is equal to $d_{j-1}c_{n-j,k-j}$. Similarly, for $j \in [n-k]$, the number of random walk paths of length $2n$ which start out negative, return to 0 for the first time at step $2j$, and are positive for exactly $2k$ steps is equal to $d_{j-1}c_{n-j,k}$. Thus, we obtain the recursion relation

$$c_{n,k} = \sum_{j=1}^{k} d_{j-1}c_{n-j,k-j} + \sum_{j=1}^{n-k} d_{j-1}c_{n-j,k}, \; k \in [n-1]. \tag{4.27}$$

Let $e_n := \binom{2n}{n}$, $n \geq 0$. As follows from the remark after Proposition 4.3, for $n \geq 1$, e_n is the number of random walk paths of length $2n$ that are equal to 0 at step $2n$. We derive a recursion formula for $\{e_n\}_{n=0}^{\infty}$. A random walk path of length $2n$ which is equal to 0 at step $2n$ must return to 0 for the first time at step $2k$, for some $k \in [n]$. The number of random walk paths of length $2n$ which are equal to 0 at time $2n$ and which return to 0 for the first time at step $2k$ is equal to $2v_k e_{n-k} = 2d_{k-1}e_{n-k}$. Consequently, we obtain the recursion formula

$$e_n = \sum_{k=1}^{n} 2d_{k-1}e_{n-k}. \tag{4.28}$$

We can now prove (4.26) by considering (4.27) and (4.28) and applying induction.

To prove (4.26) we need to show that for all $n \geq 1$,

$$c_{n,k} = e_k e_{n-k}, \text{ for } k = 0, 1, \ldots, n. \tag{4.29}$$

When $n = 1$, (4.29) clearly holds. We now assume that (4.29) holds for all $n \le n_0$ and prove that it also holds for $n = n_0 + 1$. When $n = n_0 + 1$ and $k = 0$ or $k = n_0 + 1$, we already know that (4.29) holds. So we need to show that (4.29) holds for $n = n_0 + 1$ and $k \in [n_0]$. Using (4.27) for the first equality, using the inductive assumption for the second equality, and using (4.28) for the third equality, we have

$$c_{n_0+1,k} = \sum_{j=1}^{k} d_{j-1} c_{n_0+1-j,k-j} + \sum_{j=1}^{n_0+1-k} d_{j-1} c_{n_0+1-j,k} =$$

$$\sum_{j=1}^{k} d_{j-1} e_{k-j} e_{n_0+1-k} + \sum_{j=1}^{n_0+1-k} d_{j-1} e_k e_{n_0+1-k-j} =$$

$$\frac{1}{2} e_k e_{n_0+1-k} + \frac{1}{2} e_{n_0+1-k} e_k = e_k e_{n_0+1-k}, \tag{4.30}$$

which proves that (4.29) holds for $n = n_0 + 1$ and completes the proof of Theorem 4.2. □

Exercise 4.1. This exercise completes the proof of Proposition 4.1. We proved that with probability one, the simple, symmetric random walk on $\mathbb{Z}$ visits 0 infinitely often.

(a) For fixed $x \in \mathbb{Z}$, use the fact that with probability one the random walk visits 0 infinitely often to show that with probability one the random walk visits x infinitely often. (Hint: Every time the process returns to 0, it has probability $(\frac{1}{2})^{|x|}$ of moving directly to x in $|x|$ steps.)
(b) Show that with probability one the random walk visits every $x \in \mathbb{Z}$ infinitely often.

Exercise 4.2. In this exercise, you will prove that $ET_0 = \infty$, where T_0 is the first return time to 0. We can consider the random walk starting from any $j \in \mathbb{Z}$, rather than just from 0. When we start the random walk from j, denote the corresponding probabilities and expectations by P_j and E_j. Fix $n \ge 1$ and consider starting the random walk from some $j \in \{0, 1, \ldots, n\}$. Let $T_{0,n}$ denote the first nonnegative time that the random walk is at 0 or n.

(a) Define $g(j) = E_j T_{0,n}$. By analyzing what happens on the first step, show that g solves the difference equation $g(j) = 1 + \frac{1}{2} g(j+1) + \frac{1}{2} g(j-1)$, for $j = 1, \ldots, n-1$. Note that one has the boundary conditions $g(0) = g(n) = 0$.
(b) Use (a) to show that $E_j T_{0,n} = j(n-j)$. (Hint: Write the difference equation in the form $g(j+1) - g(j) = g(j) - g(j-1) - 2$.)
(c) In particular, (b) gives $E_1 T_{0,n} = n - 1$. From this, conclude that $ET_0 = \infty$.

Exercise 4.3. Prove (4.10): $\lim_{n\to\infty} P(\frac{O^0_{2n}}{2n} > \epsilon) = 0$, for all $\epsilon > 0$. (Hint: Represent O^0_{2n} by $O^0_{2n} = \sum_{j=1}^{2n} 1_{\{S_j=0\}}$, where $1_{\{S_j=0\}}$ is as in the proof of

Proposition 4.1. From this representation, show that $\lim_{n\to\infty} E\frac{O_{2n}^0}{2n} = 0$. Conclude from this that (4.10) holds.)

Exercise 4.4. Use Stirling's formula to prove (4.24). That is, show that for any $\epsilon, \delta > 0$, there exists an $n_{\epsilon,\delta}$ such that if $n \geq n_{\epsilon,\delta}$, then

$$1 - \delta \leq \frac{\binom{2k}{k}\binom{2n-2k}{n-k}}{2^{2n}} \pi \sqrt{k(n-k)} \leq 1 + \delta,$$

for all k satisfying $\epsilon n \leq k \leq (1-\epsilon)n$.

Exercise 4.5. If one considers a simple, symmetric random walk $\{S_k\}_{k=0}^{2n}$ up to time $2n$, the probability of seeing any particular one of the 2^{2n} random walk paths of length $2n$ is equal to 2^{-2n}. Recall from the remark after Proposition 4.3 that there are $\binom{2n}{n}$ random walk paths of length $2n$ that return to 0 at time $2n$. It follows from symmetry that conditioned on $S_{2n} = 0$, the probability of seeing any particular one of the $\binom{2n}{n}$ random walks paths of length $2n$ which return to 0 at time $2n$ is equal to $\frac{1}{\binom{2n}{n}}$.

(a) Let $p \in (0,1) - \{\frac{1}{2}\}$ and consider the simple random walk on $\mathbb{Z}$ which jumps one unit to the right with probability p and one unit to the left with probability $1-p$. Denote the random walk by $\{S_n^{(p)}\}_{n=0}^{\infty}$. Consider this random walk up to time $2n$. For each particular random walk path of length $2n$, calculate the probability of seeing this path. The answer now depends on the path.

(b) Conditioned on $S_{2n}^{(p)} = 0$, show that the probability of seeing any particular one of the $\binom{2n}{n}$ random walk paths of length $2n$ which return to 0 at time $2n$ is equal to $\frac{1}{\binom{2n}{n}}$.

Exercise 4.6. Let $0 \leq j \leq m$. Consider the random walk $\{S_n^{(p)}\}_{n=0}^{\infty}$ as in Exercise 4.5, with $p \in (0,1)$, but starting from j, and denote probabilities by P_j. Let $T_{0,m}^{(p)}$ denote the first nonnegative time that this random walk is at 0 or at m. Use the method of Exercise 4.2—analyzing what happens on the first step—to calculate $P_j(S_{T_{0,m}^{(p)}}^{(p)} = 0)$, that is, the probability that starting from j, the random walk reaches 0 before it reaches m. (Hint: The calculation in the case $p = \frac{1}{2}$ needs to be treated separately.)

Chapter Notes

The arcsine law in Theorem 4.2 was first proven by P. Lévy in 1939 in the context of *Brownian motion*, which is a continuous time and continuous path version of the simple, symmetric random walk. The proof of Theorem 4.2 is due to K.L. Chung and W. Feller. One can find a proof in volume 1 of Feller's classic text in probability [19].

One can also find there a proof of Theorem 4.1. Our proofs of these theorems are a little different from Feller's proofs. As expected, the proofs in Feller's book have a probabilistic flavor. We have taken a more combinatorial/counting approach via generating functions. Proposition 4.3 and Corollary 4.2 can be derived alternatively via the "reflection principle"; see [19]. For a nice little book on random walks from the point of view of electrical networks, see Doyle and Snell [15]; for a treatise on random walks, see the book by Spitzer [32].

Chapter 5
The Distribution of Cycles in Random Permutations

In this chapter we study the limiting behavior of the total number of cycles and of the number of cycles of fixed length in random permutations of $[n]$ as $n \to \infty$. This class of problems springs from a classical question in probability called the *envelope matching* problem. You have n letters and n addressed envelopes. If you randomly place one letter in each envelope, what is the asymptotic probability as $n \to \infty$ that no letter is in its correct envelope?

Let S_n denote the set of permutations of $[n]$. Of course, S_n is a group, but the group structure will not be relevant for our purposes. For us, a permutation $\sigma \in S_n$ is simply a 1-1 map of $[n]$ onto $[n]$. The notation σ_j will be used to denote the image of $j \in [n]$ under this map. We have $|S_n| = n!$. Let P_n^U denote the uniform probability measure on S_n. That is, $P_n^U(A) = \frac{|A|}{n!}$, for any subset $A \subset S_n$. If $\sigma_j = j$, then j is called a *fixed point* for the permutation σ. Let $D_n \subset S_n$ denote the set of permutations that do not fix any points; that is, $\sigma \in D_n$ if $\sigma_j \neq j$, for all $j \in [n]$. Such permutations are called *derangements*. The classical envelope matching problem then asks for $\lim_{n\to\infty} P_n^U(D_n)$.

The standard way to solve the envelope matching problem is by the method of *inclusion–exclusion*. Define $G_i = \{\sigma \in S_n : \sigma_i = i\}$. (We suppress the dependence of G_i on n since n is fixed in this discussion.) Then the complement D_n^c of D_n is given by $D_n^c = \cup_{i=1}^n G_i$, and the inclusion–exclusion principle states that

$$P(\cup_{i=1}^n G_i) = \sum_{i=1}^n P(G_i) - \sum_{1\le i<j\le n} P(G_i \cap G_j) + \sum_{1\le i<j<k\le n} P(G_i \cap G_j \cap G_k) - \cdots + (-1)^{n-1} P(\cap_{i=1}^n G_i).$$

(See Exercise A.2 in Appendix A.) Each of the probabilities above can be computed readily. After some calculations one finds that $P(D_n) = 1 - P(\cup_{i=1}^n G_i) = 1 - 1 + \frac{1}{2!} - \frac{1}{3!} + \cdots + (-1)^n \frac{1}{n!}$; thus, $\lim_{n\to\infty} P(D_n) = e^{-1}$.

R.G. Pinsky, *Problems from the Discrete to the Continuous*, Universitext,
DOI 10.1007/978-3-319-07965-3_5, © Springer International Publishing Switzerland 2014

Here is an elegant, alternative proof using generating functions. Let $d_k^{(n)}$ denote the number of permutations in S_n that fix exactly k points. We need to calculate $\lim_{n\to\infty} \frac{d_0^{(n)}}{n!}$. Clearly,

$$\sum_{k=0}^{n} d_k^{(n)} = n!, \tag{5.1}$$

since every permutation fixes k points, for some k. To construct a permutation in S_n that fixes exactly k points, first we can choose k numbers from $[n]$ for the fixed points, and then we must choose a permutation of the other $n-k$ numbers that fixes none of them; thus,

$$d_k^{(n)} = \binom{n}{k} d_0^{(n-k)}.$$

Substituting this in (5.1) gives

$$\sum_{k=0}^{n} \binom{n}{k} d_0^{(n-k)} = n!,$$

or equivalently

$$\sum_{k=0}^{n} \frac{d_0^{(n-k)}}{k!(n-k)!} = 1. \tag{5.2}$$

If one multiplies the absolutely convergent power series $\sum_{n=0}^{\infty} a_n x^n$ by the absolutely convergent power series $\sum_{n=0}^{\infty} b_n x^n$, one gets the absolutely convergent power series $\sum_{n=0}^{\infty} c_n x^n$, where $c_n = \sum_{k=0}^{n} a_k b_{n-k}$. Thus, it follows from (5.2) that

$$\Big(\sum_{n=0}^{\infty} \frac{x^n}{n!}\Big)\Big(\sum_{n=0}^{\infty} \frac{d_0^{(n)}}{n!} x^n\Big) = \sum_{n=0}^{\infty} x^n, \ |x| < 1,$$

or

$$\sum_{n=0}^{\infty} \frac{d_0^{(n)}}{n!} x^n = \frac{e^{-x}}{1-x}, \ |x| < 1. \tag{5.3}$$

Thus $\frac{d_0^{(n)}}{n!}$ is the coefficient of x^n in

$$\frac{e^{-x}}{1-x} = \Big(1 - x + \frac{x^2}{2!} - \frac{x^3}{3!} + \cdots\Big)(1 + x + x^2 + x^3 + \cdots),$$

and this is easily seen to give $\frac{d_0^{(n)}}{n!} = 1 - 1 + \frac{1}{2!} - \frac{1}{3!} + \cdots + (-1)^n \frac{1}{n!}$.

In order to begin our study of the behavior of the number of cycles and of the number of cycles of fixed length in random permutations, we recall some basic facts and notation concerning cycles of permutations. Consider the permutation $\sigma \in S_4$ given in two-line form by $\binom{1\,2\,3\,4}{2\,4\,1\,3}$. This means that $\sigma_1 = 2, \sigma_2 = 4$, etc. Since 1 goes to 2, 2 goes to 4, 4 goes to 3, and 3 goes back to 1, we call σ *cyclic* and denote this by writing $\sigma = (1\,2\,4\,3)$. (We could also just as well write it as $(4\,3\,1\,2)$, for example.) Recall that every permutation can be decomposed into a product of disjoint cycles. For example, consider $\sigma \in S_8$ given by $\binom{1\,2\,3\,4\,5\,6\,7\,8}{3\,2\,5\,8\,6\,7\,1\,4}$. Under σ, 1 goes to 3, 3 goes to 5, 5 goes to 6, 6 goes to 7, and 7 goes back to 1, closing a cycle. Now 2 goes to 2, which makes a cycle unto itself, and finally, 4 goes to 8 and 8 goes back to 4. Therefore, we write $\sigma = (1\,3\,5\,6\,7)(2)(4\,8)$ or, alternatively, $\sigma = (1\,3\,5\,6\,7)(4\,8)$; in the latter form, the convention is that every number that does not appear at all forms a cycle unto itself. Note that σ has one cycle of length 5, one cycle of length 2, and one cycle of length 1.

For $\sigma \in S_n$ and $j \in [n]$, let $C_j^{(n)}(\sigma)$ denote the number of cycles of length j in σ. Note that for all $\sigma \in S_n$, one has the identity

$$\sum_{j=1}^{n} jC_j^{(n)}(\sigma) = n.$$

We call $(C_1^{(n)}(\sigma), C_2^{(n)}(\sigma), \ldots, C_n^{(n)}(\sigma))$ the *cycle type* of the permutation σ. Let

$$N^{(n)}(\sigma) = \sum_{j=1}^{n} C_j^{(n)}(\sigma)$$

denote the number of cycles in the permutation $\sigma \in S_n$. Under the probability measure P_n^U, we may think of $N^{(n)}$ and $C_j^{(n)}$ as random variables. In this chapter we will investigate the limiting distribution of the random variable $N^{(n)}$ and of the random variable $C_j^{(n)}$ for fixed j, as $n \to \infty$. In fact, more generally, we will investigate the limiting distribution of the j-dimensional random vector $(C_1^{(n)}, C_2^{(n)}, \ldots, C_j^{(n)})$. We call these cycles *small cycles* because their lengths are fixed as $n \to \infty$.

Instead of just considering permutations under the uniform measure P_n^U, we will consider permutations under a one-parameter family of probability measures which includes the uniform measure as a particular case. For each $\theta \in (0, \infty)$, we define a probability measure $P_n^{(\theta)}$ on S_n by

$$P_n^{(\theta)}(\{\sigma\}) = \frac{\theta^{N^{(n)}(\sigma)}}{K_n(\theta)},$$

where

$$K_n(\theta) = \sum_{\sigma \in S_n} \theta^{N^{(n)}(\sigma)}$$

is the normalizing constant required to make a probability measure. Thus, under the measure $P_n^{(\theta)}$, every permutation is weighted proportionally by the parameter θ raised to an exponent equal to the number of cycles in the permutation. Consequently, for $\theta > 1$, $P_n^{(\theta)}$ favors permutations with many cycles, and for $\theta < 1$, it favors permutations with few cycles. Of course $\theta = 1$ corresponds to the uniform measure: $P_n^U = P_n^{(1)}$. The original reason for considering the probability measures $P_n^{(\theta)}$ can be attributed to Proposition 5.1 below, which gives the exact distribution of the cycle types under $P_n^{(\theta)}$. In Exercise 5.1, the reader is asked to verify that Proposition 5.1 follows from the definition of $P_n^{(\theta)}$ along with Proposition 5.2 and Lemma 5.1, which are stated and proved in the course of the proofs of Theorems 5.1 and 5.2 below. We use the standard notation

$$\theta^{(n)} := \theta(\theta + 1) \cdots (\theta + n - 1),\ n \geq 1.$$

This expression is sometimes referred to as a *rising factorial*; the notation is called the *Pochhammer* symbol.

Proposition 5.1. *If* $\sum_{j=1}^{n} ja_j = n$, *then*

$$P_n^{(\theta)}(C_1^{(n)} = a_1, C_2^{(n)} = a_2, \ldots, C_n^{(n)} = a_n) = \frac{n!}{\theta^{(n)}} \prod_{j=1}^{n} (\frac{\theta}{j})^{a_j} \frac{1}{a_j!}.$$

The distribution in Proposition 5.1 is known as the *Ewens sampling formula*; it arose originally in the context of population genetics.

We will prove a *weak law of large numbers* for the distribution of the number of cycles $N^{(n)}$.

Theorem 5.1. *Let* $\theta \in (0, \infty)$. *Under* $P_n^{(\theta)}$, *the distribution of the number of cycles* $N^{(n)}$ *in a permutation satisfies*

$$\frac{N^{(n)}}{\log n} \to \theta \text{ in probability;}$$

that is, for all $\epsilon > 0$,

$$\lim_{n \to \infty} P_n^{(\theta)}(|\frac{N^{(n)}}{\log n} - \theta| \geq \epsilon) = 0.$$

We now consider the small cycles. A random variable Z is distributed according to the Poisson distribution with parameter $\lambda > 0$ ($Z \sim \text{Pois}(\lambda)$) if

$$P(Z = j) = e^{-\lambda}\frac{\lambda^j}{j!}, \text{ for } j = 0, 1, \ldots.$$

The j discrete random variables $\{X_i\}_{i=1}^j$ are called independent if $P(X_1 = x_1, \ldots, X_j = x_j) = \prod_{i=1}^j P(X_i = x_i)$, for all choices of $\{x_i\}_{i=1}^j \subset \mathbb{R}$. In the sequel, Z_λ will denote a random variable distributed according to $\text{Pois}(\lambda)$, and it will always be assumed that $\{Z_{\lambda_i}\}_{i=1}^j$ are independent for distinct $\{\lambda_i\}_{i=1}^j$.

We will prove a *weak convergence* result for small cycles.

Theorem 5.2. *Let $\theta \in (0, \infty)$. Let j be a positive integer. Under the measure $P_n^{(\theta)}$, the distribution of the random vector $(C_1^{(n)}, C_2^{(n)}, \ldots, C_j^{(n)})$ converges weakly to the distribution of $(Z_\theta, Z_{\frac{\theta}{2}}, \ldots, Z_{\frac{\theta}{j}})$. That is,*

$$\lim_{n\to\infty} P_n^{(\theta)}(C_1^{(n)} = m_1, C_2^{(n)} = m_2, \ldots, C_j^{(n)} = m_j) = \prod_{i=1}^j e^{-\frac{\theta}{i}}\frac{(\frac{\theta}{i})^{m_i}}{m_i!},$$
$$m_i \geq 0,\ i = 1, \ldots, j. \tag{5.4}$$

Remark. Let j be a positive integer and let $1 \leq k_1 < k_2 < \cdots < k_j$. In Exercise 5.7 the reader is asked to show that by making a small change in the proof of Theorem 5.2, one has

$$\lim_{n\to\infty} P_n^{(\theta)}(C_{k_1}^{(n)} = m_1, C_{k_2}^{(n)} = m_2, \ldots, C_{k_j}^{(n)} = m_j) = \prod_{i=1}^j e^{-\frac{\theta}{k_i}}\frac{(\frac{\theta}{k_i})^{m_i}}{m_i!},$$
$$m_i \geq 0,\ i = 1, \ldots, j. \tag{5.5}$$

In particular, for any fixed j, the distribution of $C_j^{(n)}$ converges weakly to the $\text{Pois}(\frac{\theta}{j})$ distribution. Actually, (5.5) can be deduced directly from (5.4); see Exercises 5.2 and 5.3.

Our proofs of these two theorems will be very combinatorial, through the method of generating functions. The use of purely probabilistic reasoning will be rather minimal.

For the proofs of the two theorems, we will need to evaluate the normalizing constant $K_n(\theta)$. Of course, this is trivial in the case of the uniform measure, that is, the case $\theta = 1$. Let $s(n,k)$ denote the number of permutations in S_n that have exactly k cycles. From the definition of $K_n(\theta)$, we have

$$K_n(\theta) = \sum_{k=1}^n s(n,k)\theta^k. \tag{5.6}$$

Proposition 5.2.

$$K_n(\theta) = \theta^{(n)}.$$

Remark. The numbers $s(n,k)$ are called *unsigned Stirling numbers of the first kind*. Proposition 5.2 and (5.6) show that they arise as the coefficients of the polynomials $q_n(\theta) := \theta^{(n)} = \theta(\theta+1)\cdots(\theta+n-1)$.

Proof. There are $(n-1)!$ permutations in S_n that contain only one cycle and one permutation in S_n that contains n cycles:

$$s(n,1) = (n-1)!, \;\; s(n,n) = 1. \tag{5.7}$$

We prove the following recursion relation:

$$s(n+1,k) = ns(n,k) + s(n,k-1), \; n \geq 2, \; 2 \leq k \leq n. \tag{5.8}$$

Note that (5.7) and (5.8) uniquely determine $s(n,k)$ for all $n \geq 1$ and all $k \in [n]$.

To create a permutation $\sigma' \in S_{n+1}$, we can start with a permutation $\sigma \in S_n$ and then take the number $n+1$ and either insert it into one of the existing cycles of σ or let it stand alone as a cycle of its own. If we insert $n+1$ into one of the existing cycles, then σ' will have k cycles if and only if σ has k cycles. There are n possible locations in which one can place the number $n+1$ and preserve the number of cycles. (The reader should verify this.) Thus, from each permutation in S_n with k cycles, we can construct n permutations in S_{n+1} with k cycles. If, on the other hand, we let $n+1$ stand alone in its own cycle, then σ' will have k cycles if and only if σ has $k-1$ cycles. Thus, from each permutation in S_n with $k-1$ cycles, we can construct one permutation in S_{n+1} with k cycles. Now (5.8) is the mathematical expression of this verbal description.

Let $c_{n,k}$ denote the coefficient of θ^k in $q_n(\theta) = \theta(\theta+1)\cdots(\theta+n-1)$. Clearly $c_{n,1} = (n-1)!$ and $c_{n,n} = 1$, for $n \geq 1$. Writing $q_{n+1}(\theta) = q_n(\theta)(\theta+n)$, one sees that $c_{n+1,k} = nc_{n,k} + c_{n,k-1}$, for $n \geq 2, \; 2 \leq k \leq n$. Thus, $c_{n,k}$ satisfies the same recursion relation (5.8) and the same boundary condition (5.7) as does $s(n,k)$. We conclude that $c_{n,k} = s(n,k)$. The proposition follows from this along with (5.6). □

In light of Proposition 5.2, from now on, we write the probability measure $P_n^{(\theta)}$ in the form

$$P_n^{(\theta)}(\{\sigma\}) = \frac{\theta^{N^{(n)}(\sigma)}}{\theta^{(n)}}.$$

We now set the stage to prove Theorem 5.1. The *probability generating function* $P_X(s)$ of a random variable X taking nonnegative integral values is defined by

$$P_X(s) = Es^X = \sum_{i=0}^{\infty} s^i P(X=i), \; |s| \leq 1.$$

The probability generating function uniquely determines the distribution; indeed, $\frac{1}{i!}\frac{d^i P_X(s)}{ds^i}|_{s=0} = P(X = i)$. Let $P_{N^{(n)}}(s;\theta)$ denote the probability generating function for the random variable $N^{(n)}$ under $P_n^{(\theta)}$:

$$P_{N^{(n)}}(s;\theta) = \sum_{i=1}^{n} s^i P_n^{(\theta)}(N^{(n)} = i).$$

Recalling that $s(n,i)$ denotes the number of permutations in S_n with i cycles, it follows that

$$P_n^{(\theta)}(N^{(n)} = i) = \frac{\theta^i s(n,i)}{\theta^{(n)}}.$$

Using this with (5.6) and Proposition 5.2 gives

$$P_{N^{(n)}}(s;\theta) = \sum_{i=1}^{n} s^i \frac{\theta^i s(n,i)}{\theta^{(n)}} = \frac{(s\theta)^{(n)}}{\theta^{(n)}} = \frac{s\theta(s\theta+1)\cdots(s\theta+n-1)}{\theta(\theta+1)\cdots(\theta+n-1)} =$$

$$\prod_{i=1}^{n}\left(\frac{\theta}{\theta+i-1}s + \frac{i-1}{\theta+i-1}\right). \tag{5.9}$$

A random variable X is distributed according to the Bernoulli distribution with parameter $p \in [0,1]$ if $P(X = 1) = p$ and $P(X = 0) = 1 - p$. We write $X \sim \text{Ber}(p)$. The probability generating function for such a random variable is $ps + 1 - p$. Now let $\{X_{\theta(\theta+i-1)^{-1}}\}_{i=1}^n$ be independent random variables, where $X_{\theta(\theta+i-1)^{-1}} \sim \text{Ber}(\frac{\theta}{\theta+i-1})$. Let $Z_{n,\theta} = \sum_{i=1}^n X_{\theta(\theta+i-1)^{-1}}$. Then the probability generating function for $Z_{n,\theta}$ is given by

$$P_{Z_{n,\theta}}(s) = Es^{Z_{n,\theta}} = Es^{\sum_{i=1}^n X_{\theta(\theta+i-1)^{-1}}} = \prod_{i=1}^{n} Es^{X_{\theta(\theta+i-1)^{-1}}} =$$

$$\prod_{i=1}^{n}\left(\frac{\theta}{\theta+i-1}s + \frac{i-1}{\theta+i-1}\right). \tag{5.10}$$

For the third equality above we have used the fact that the expected value of a product of independent random variables is equal to the product of their expected values. From (5.9), (5.10), and the uniqueness of the probability generating function, we obtain the following proposition.

Proposition 5.3. *Under $P_n^{(\theta)}$, the distribution of $N^{(n)}$ is equal to the distribution of $\sum_{i=1}^n X_{\theta(\theta+i-1)^{-1}}$, where $\{X_{\theta(\theta+i-1)^{-1}}\}_{i=1}^n$ are independent random variables, and $X_{\theta(\theta+i-1)^{-1}} \sim Ber(\frac{\theta}{\theta+i-1})$.*

Remark. As an alternative way of arriving at the result in the proposition, there is a nice probabilistic construction of uniformly random permutations ($\theta = 1$) that immediately yields the result, and the construction can be amended to cover the case of general θ. See Exercise 5.4.

We now use Proposition 5.3 and Chebyshev's inequality to prove the first theorem.

Proof of Theorem 5.1. Let $Z_{n,\theta} = \sum_{i=1}^{n} X_{\theta(\theta+i-1)^{-1}}$. By Proposition 5.3, it suffices to show that

$$\lim_{n\to\infty} P(|\frac{Z_{n,\theta}}{\log n} - \theta| \ge \epsilon) = 0, \text{ for all } \epsilon > 0. \tag{5.11}$$

If $X_p \sim \text{Ber}(p)$, then the expected value of X_p is $EX_p = p$, and the variance is $\text{Var}(X_p) = p(1-p)$. Since the expectation is linear, we have $EZ_{n,\theta} = \sum_{i=1}^{n} \frac{\theta}{\theta+i-1}$. By considering the above sum simultaneously as an upper Riemann sum and as a lower Riemann sum of appropriate integrals, we have

$$\theta\big(\log(n+\theta) - \log\theta\big) = \theta\int_0^n \frac{1}{\theta+x}\,dx \le \sum_{i=1}^{n} \frac{\theta}{\theta+i-1} = EZ_{n,\theta} \le$$

$$1 + \theta\int_0^{n-1} \frac{1}{\theta+x}\,dx = 1 + \theta\big(\log(n-1+\theta) - \log\theta\big).$$

Since $\log(n+\theta) = \log n + \log(1+\frac{\theta}{n})$ and $\log(n-1+\theta) = \log n + \log(1+\frac{\theta-1}{n})$, the above inequality immediately yields

$$EZ_{n,\theta} = \theta\log n + O(1), \text{ as } n \to \infty. \tag{5.12}$$

Since the variance of a sum of independent random variables is the sum of the variances of the random variables, we have $\text{Var}(Z_{n,\theta}) = \sum_{i=1}^{n} \frac{\theta(i-1)}{(\theta+i-1)^2}$. Similar to the integral estimate above for the expectation, we have

$$\text{Var}(Z_{n,\theta}) \le \theta\sum_{i=2}^{n} \frac{1}{i-1} \le \theta + \theta\int_1^{n-1} \frac{1}{x}\,dx = \theta + \theta\log(n-1). \tag{5.13}$$

Using (5.12) for the last inequality below, we have for sufficiently large n

$$\begin{aligned}
&P(|\frac{Z_{n,\theta}}{\log n} - \theta| \ge \epsilon) = P(|Z_{n,\theta} - \theta\log n| \ge \epsilon\log n) = \\
&P(|(Z_{n,\theta} - EZ_{n,\theta}) + (EZ_{n,\theta} - \theta\log n)| \ge \epsilon\log n) \le \\
&P(|Z_{n,\theta} - EZ_{n,\theta}| \ge \epsilon\log n - |EZ_{n,\theta} - \theta\log n)|) \le \\
&P(|Z_{n,\theta} - EZ_{n,\theta}| \ge \frac{1}{2}\epsilon\log n).
\end{aligned} \tag{5.14}$$

Applying Chebyshev's inequality to the last term in (5.14), it follows from (5.13) and (5.14) that for sufficiently large n,

$$P(|\frac{Z_{n,\theta}}{\log n} - \theta| \geq \epsilon) \leq \frac{\theta + \theta\log(n-1)}{\frac{1}{4}\epsilon^2\log^2 n}. \tag{5.15}$$

Now (5.11) follows from (5.15). □

We now develop a framework that will lead to the proof of Theorem 5.2. Given a positive integer n and given a collection $\{a_i\}_{i=1}^n$ of nonnegative integers satisfying $\sum_{i=1}^n ia_i = n$, let $c_n(a_1,\ldots,a_n)$ denote the number of permutations $\sigma \in S_n$ with cycle type $(a_1,\ldots,a_n)$. From the definition of $P_n^{(\theta)}$, we have

$$P_n^{(\theta)}(C_1^{(n)} = a_1, C_2^{(n)} = a_2, \ldots, C_n^{(n)} = a_n) = \frac{\theta^{\sum_{i=1}^n a_i} c_n(a_1,\ldots,a_n)}{\theta^{(n)}}.$$

To prove Theorem 5.2, we need to analyze $P_n^{(\theta)}(C_1^{(n)} = m_1,\ldots,C_j^{(n)} = m_j)$, for large n and fixed j. We have

$$P_n^{(\theta)}(C_1^{(n)} = m_1, C_2^{(n)} = m_2, \ldots, C_j^{(n)} = m_j) =$$

$$\sum_{\substack{\sum_{i=1}^j im_i + \sum_{i=j+1}^n ia_i = n \\ a_{j+1}\geq 0,\ldots,a_n\geq 0}} P_n^{(\theta)}(C_1^{(n)}=m_1,\ldots,C_j^{(n)}=m_j, C_{j+1}^{(n)}=a_{j+1},\ldots,C_n^{(n)}=a_n)=$$

$$\sum_{\substack{\sum_{i=1}^j im_i + \sum_{i=j+1}^n ia_i = n \\ a_{j+1}\geq 0,\ldots,a_n\geq 0}} \frac{\theta^{\sum_{i=1}^j m_i + \sum_{i=j+1}^n a_i} c_n(m_1,\ldots,m_j,a_{j+1},\ldots,a_n)}{\theta^{(n)}}. \tag{5.16}$$

We calculate $c_n(a_1,\ldots,a_n)$ by direct combinatorial reasoning.

Lemma 5.1.

$$c_n(a_1,\ldots,a_n) = \frac{n!}{\prod_{i=1}^n i^{a_i} a_i!}.$$

Remark. From the lemma and (5.16), we obtain

$$P_n^{(\theta)}(C_1^{(n)} = m_1, C_2^{(n)} = m_2, \ldots, C_j^{(n)} = m_j) =$$

$$\frac{n!}{\theta^{(n)}} \prod_{i=1}^j \frac{(\frac{\theta}{i})^{m_i}}{m_i!} \sum_{\substack{\sum_{i=1}^j im_i + \sum_{i=j+1}^n ia_i = n \\ a_{j+1}\geq 0,\ldots,a_n\geq 0}} \prod_{i=j+1}^n \frac{(\frac{\theta}{i})^{a_i}}{a_i!}.$$

The sum on the right hand side above is a real mess; however, a sophisticated application of generating functions in conjunction with the lemma will allow us to evaluate the right hand side of (5.16) indirectly.

Proof of Lemma 5.1. First we separate out a_1 numbers for 1 cycles, $2a_2$ numbers for 2 cycles,..., $(n-1)a_{n-1}$ numbers for $(n-1)$ cycles, and finally the last na_n numbers for n cycles. The number of ways of doing this is

$$\binom{n}{a_1}\binom{n-a_1}{2a_2}\binom{n-a_1-2a_2}{3a_3}\cdots\binom{n-a_1-\cdots-(n-1)a_{n-1}}{na_n}=$$
$$\frac{n!}{a_1!(2a_2)!\cdots(na_n)!}.$$

The a_1 numbers selected for 1 cycles need no further differentiation. The $2a_2$ numbers selected for 2 cycles must be separated out into a_2 pairs. Of course the order of the pairs is irrelevant, so the number of ways of doing this is

$$\frac{1}{a_2!}\binom{2a_2}{2}\binom{2a_2-2}{2}\cdots\binom{4}{2}\binom{2}{2}=\frac{(2a_2)!}{a_2!(2!)^{a_2}}.$$

The $3a_3$ numbers selected for 3 cycles must be separated out into a_3 triplets, and then each such triplet must be ordered in a cycle. The number of ways of separating the $3a_3$ numbers into triplets is

$$\frac{1}{a_3!}\binom{3a_2}{3}\binom{3a_3-3}{3}\cdots\binom{6}{3}\binom{3}{3}=\frac{(3a_3)!}{a_3!(3!)^{a_3}}.$$

Each such triplet can be ordered into a cycle in $(3-1)!$ ways. Thus, we conclude that the $3a_3$ numbers can be arranged into a_3 3 cycles in $\frac{((3-1)!)^{a_3}(3a_3)!}{a_3!(3!)^{a_3}}$ ways. Continuing like this, we obtain

$$c(a_1,\ldots,a_n)=\frac{n!}{a_1!(2a_2)!\cdots(na_n)!}\frac{(2a_2)!}{a_2!(2!)^{a_2}}\frac{((3-1)!)^{a_3}(3a_3)!}{a_3!(3!)^{a_3}}\cdots\frac{((n-1)!)^{a_n}(na_n)!}{a_n!(n!)^{a_n}}=$$
$$\frac{n!}{a_1!a_2!\cdots a_n!2^{a_2}3^{a_3}\cdots n^{a_n}}.$$

□

We now turn to generating functions. Consider an infinite dimensional vector $x=(x_1,x_2,\ldots)$, and for any positive integer n, define $x^{(n)}=(x_1,\ldots,x_n)$. For $a=(a_1,\ldots,a_n)$, let $x^a=(x^{(n)})^a:=x_1^{a_1}\cdots x_n^{a_n}$. Let $T(\sigma)$ denote the cycle type of $\sigma\in S_n$. Define the *cycle index* of S_n, $n\geq 1$, by

$$\phi_n(x) = \phi_n(x^{(n)}) = \frac{1}{n!}\sum_{\sigma \in G} x^{T(\sigma)} = \frac{1}{n!}\sum_{\substack{\sum_{i=1}^n ia_i = n \\ a_1 \geq 0, \ldots, a_n \geq 0}} c_n(a)x^a.$$

We also define $\phi_0(x) = 1$. We now consider (formally for the moment) the generating function for $\phi_n(\theta x)$:

$$G^{(\theta)}(x,t) = \sum_{n=0}^{\infty} \phi_n(\theta x)t^n, \quad x = (x_1, x_2, \ldots).$$

Using Lemma 5.1, we can obtain a very nice representation for $G^{(\theta)}$, as well as a domain on which its defining series converges. Let $||x||_\infty := \sup_{n\geq 1} |x_n|$.

Proposition 5.4.

$$G^{(\theta)}(x,t) = \exp(\theta \sum_{i=1}^{\infty} \frac{x_i t^i}{i}), \ \textit{for } |t| < 1, \ ||x||_\infty < \infty.$$

Proof. Consider $t \in [0,1)$ and x with $x_j \geq 0$ for all j, and $||x||_\infty < \infty$. Using Lemma 5.1 and the definition of $\phi_n(x)$, we have

$$G^{(\theta)}(x,t) = \sum_{n=0}^{\infty} \sum_{\substack{\sum_{j=1}^n ja_j = n \\ a_1 \geq 0, \ldots, a_n \geq 0}} \frac{c_n(a)(\theta x)^a t^n}{n!} =$$

$$\sum_{n=0}^{\infty} \sum_{\substack{\sum_{j=1}^n ja_j = n \\ a_1 \geq 0, \ldots, a_n \geq 0}} \frac{n!}{\prod_{i=1}^n i^{a_i} a_i!} \frac{(\theta x_1)^{a_1} \cdots (\theta x_n)^{a_n} t^n}{n!} =$$

$$\sum_{n=0}^{\infty} \sum_{\substack{\sum_{j=1}^n ja_j = n \\ a_1 \geq 0, \ldots, a_n \geq 0}} \prod_{i=1}^{n} \frac{(\frac{\theta x_i t^i}{i})^{a_i}}{a_i!} = \sum_{a_1 \geq 0, a_2 \geq 0, \ldots} \prod_{i=1}^{\infty} \frac{(\frac{\theta x_i t^i}{i})^{a_i}}{a_i!} = \prod_{i=1}^{\infty} e^{\frac{\theta x_i t^i}{i}} =$$

$$\exp(\theta \sum_{i=1}^{\infty} \frac{x_i t^i}{i}). \tag{5.17}$$

The right hand side above converges for t and x in the range specified at the beginning of the proof. Since all of the summands in sight are nonnegative, it follows that the series defining $G^{(\theta)}$ is convergent in this range. For t and x in the range specified in the statement of the theorem, the above calculation shows that there is absolute convergence and hence convergence. □

We now exploit the formula for $G^{(\theta)}(x,t)$ in Proposition 5.4 in a clever way. Recall that

$$\log(1-t) = -\sum_{i=1}^{\infty} \frac{t^i}{i}. \tag{5.18}$$

For $x = (x_1, x_2, \ldots)$ and a positive integer j, let $x^{(j);1} = (x_1, \ldots, x_j, 1, 1, \ldots)$. In other words, $x^{(j);1}$ is the infinite dimensional vector which coincides with x in its first j places and has 1 in all of its other places. From Proposition 5.4 and (5.18) we have

$$G^{(\theta)}(x^{(j);1}, t) = \exp(\theta \sum_{i=1}^{\infty} \frac{t^i}{i}) \exp(\theta \sum_{i=1}^{j} \frac{(x_i-1)t^i}{i}) = \frac{1}{(1-t)^\theta} \exp(\theta \sum_{i=1}^{j} \frac{(x_i-1)t^i}{i}). \tag{5.19}$$

We will need the following lemma.

Lemma 5.2. *Let $\theta \in (0, \infty)$. Let $\sum_{i=0}^{\infty} b_i$ be a convergent series, and assume that*

$$\frac{1}{(1-t)^\theta} \sum_{i=0}^{\infty} b_i t^i = \sum_{i=0}^{\infty} \gamma_i t^i, \ |t| < 1.$$

If $\theta > 1$, also assume that $\sum_{i=0}^{\infty} |b_i| < \infty$. If $\theta \in (0, 1)$, also assume that $\sum_{i=0}^{\infty} s^i |b_i| < \infty$, for some $s > 1$. Then

$$\lim_{n\to\infty} \frac{n!}{\theta^{(n)}} \gamma_n = \sum_{i=0}^{\infty} b_i.$$

Proof. Since $(\frac{1}{(1-t)^\theta})^{(n)}|_{t=0} = \theta(\theta+1)\cdots(\theta+n-1) = \theta^{(n)}$, the Taylor expansion for $\frac{1}{(1-t)^\theta}$ is given by

$$\frac{1}{(1-t)^\theta} = \sum_{n=0}^{\infty} \frac{\theta^{(n)}}{n!} t^n, \tag{5.20}$$

where for convenience we have defined $\theta^{(0)} = 1$. Thus, the Taylor expansion for $\frac{1}{(1-t)^\theta} \sum_{i=0}^{\infty} b_i t^i$ is given by

$$\frac{1}{(1-t)^\theta} \sum_{i=0}^{\infty} b_i t^i = \sum_{n=0}^{\infty} d_n t^n,$$

where $d_n = \sum_{i=0}^{n} b_i \frac{\theta^{(n-i)}}{(n-i)!}$. Therefore, by the assumption in the lemma, we have

$$\gamma_n = \sum_{i=0}^{n} b_i \frac{\theta^{(n-i)}}{(n-i)!}.$$

If $\theta = 1$, then $k! = \theta^{(k)}$, for all k. Consequently the above equation reduces to $\gamma_n = \sum_{i=0}^{n} b_i$, and thus the statement of the lemma holds. When $\theta \neq 1$, then using the additional assumptions on $\{b_i\}_{i=0}^{\infty}$, we can show that

$$\lim_{n\to\infty} \frac{n!}{\theta^{(n)}} \sum_{i=0}^{n} b_i \frac{\theta^{(n-i)}}{(n-i)!} = \sum_{i=0}^{\infty} b_i, \tag{5.21}$$

which finishes the proof of the lemma. The reader is guided through a proof of (5.21) in Exercise 5.5. □

We can now give the proof of Theorem 5.2.

Proof of Theorem 5.2. From (5.19) and the original definition of $G^{(\theta)}(x,t)$, we have

$$\frac{1}{(1-t)^{\theta}} \exp(\theta \sum_{i=1}^{j} \frac{(x_i - 1)t^i}{i}) = \sum_{n=0}^{\infty} \phi_n(\theta x^{(j);1}) t^n. \tag{5.22}$$

Considering x and θ as constants, we apply Lemma 5.2 to (5.22). In terms of the lemma, we have

$$\gamma_n = \phi_n(\theta x^{(j);1}) \text{ and } \exp(\theta \sum_{i=1}^{j} \frac{(x_i - 1)t^i}{i}) = \sum_{i=0}^{\infty} b_i t^i. \tag{5.23}$$

In order to be able to apply the lemma for all $\theta > 0$, we need to show that $\sum_{i=0}^{\infty} s^i |b_i| < \infty$, for some $s > 1$. Define $\{\bar{b}_i\}_{i=0}^{\infty}$ by

$$\exp(\theta \sum_{i=1}^{j} \frac{|x_i - 1| t^i}{i}) = \sum_{i=0}^{\infty} \bar{b}_i t^i. \tag{5.24}$$

Since all of the coefficients in the sum in the exponent on the left hand side of (5.24) are nonnegative, we have $\bar{b}_i \geq |b_i| \geq 0$, for all i. The reader is asked to prove this in Exercise 5.6. The function on the left hand side of (5.24) is real analytic for all $t \in \mathbb{R}$ (and complex analytic for all complex t); consequently, its power series on the right hand side converges for all $t \in \mathbb{R}$. From this and the nonnegativity of $\bar{b}_i$, it follows that $\sum_{i=0}^{\infty} s^i \bar{b}_i < \infty$, for all $s \geq 0$, and then, since $|b_i| \leq \bar{b}_i$, we conclude that $\sum_{i=0}^{\infty} s^i |b_i| < \infty$, for all $s \geq 0$.

By definition, from (5.23), we have

$$\sum_{i=0}^{\infty} b_i = \exp(\theta \sum_{i=1}^{j} \frac{x_i - 1}{i}). \tag{5.25}$$

Consider now

$$\frac{n!}{\theta^{(n)}}\gamma_n = \frac{n!}{\theta^{(n)}}\phi_n(\theta x^{(j);1}) = \frac{1}{\theta^{(n)}} \sum_{\substack{\sum_{i=1}^n ia_i = n\\ a_1\geq 0,\ldots,a_n\geq 0}} c_n(a)(\theta x^{(j);1})^a. \tag{5.26}$$

For any given j-vector $(m_1,\ldots,m_j)$ with nonnegative integral entries, the coefficient of $x_1^{m_1}x_2^{m_2}\cdots x_j^{m_j}$ in (5.26) is

$$\frac{1}{\theta^{(n)}} \sum_{\substack{\sum_{i=1}^j im_i+\sum_{i=j+1}^n ia_i = n\\ a_{j+1}\geq 0,\ldots,a_n\geq 0}} \theta^{\sum_{i=1}^j m_i + \sum_{i=j+1}^n a_i} c_n(m_1,\ldots,m_j,a_{j+1},\ldots,a_n).$$

But by (5.16), this is exactly $P_n^{(\theta)}(C_1^{(n)} = m_1, C_2^{(n)} = m_2,\ldots,C_j^{(n)} = m_j)$. By Lemma 5.2, $\lim_{n\to\infty}\frac{n!}{\theta^{(n)}}\gamma_n$ exists, and this is true for every choice of x and θ; thus, we conclude that

$$\lim_{n\to\infty}\frac{n!}{\theta^{(n)}}\gamma_n = \lim_{n\to\infty}\frac{n!}{\theta^{(n)}}\phi_n(\theta x^{(j);1}) = \sum_{m_1\geq 0,\ldots,m_j\geq 0} p_{m_1,\ldots,m_j}(\theta)x_1^{m_1}\cdots x_j^{m_j}, \tag{5.27}$$

where

$$p_{m_1,\ldots,m_j}(\theta) = \lim_{n\to\infty} P_n^{(\theta)}(C_1^{(n)} = m_1, C_2^{(n)} = m_2,\ldots,C_j^{(n)} = m_j). \tag{5.28}$$

Applying Lemma 5.2, we conclude from (5.25) and (5.27) that

$$\exp(\theta\sum_{i=1}^j \frac{x_i - 1}{i}) = \sum_{m_1\geq 0,\ldots,m_j\geq 0} p_{m_1,\ldots,m_j}(\theta)x_1^{m_1}\cdots x_j^{m_j}. \tag{5.29}$$

On the one hand, (5.29) shows that the coefficient of $x_1^{m_1}\cdots x_j^{m_j}$ in the Taylor expansion about $x = 0$ of the function $\exp(\theta\sum_{i=1}^j\frac{x_i-1}{i})$ is $p_{m_1,\ldots,m_j}(\theta)$. On the other hand, by Taylor's formula, this coefficient is equal to

$$\frac{1}{m_1!\cdots m_j!}\frac{\partial^{m_1+\cdots+m_j}\left(\exp(\theta\sum_{i=1}^j\frac{x_i-1}{i})\right)}{\partial_{x_1}^{m_1}\cdots\partial_{x_j}^{m_j}}|_{x=0} =$$

$$\frac{1}{m_1!\cdots m_j!}\exp(-\theta\sum_{i=1}^j\frac{1}{i})\prod_{i=1}^j(\frac{\theta}{i})^{m_i} = \prod_{i=1}^j e^{-\frac{\theta}{i}}\frac{(\frac{\theta}{i})^{m_i}}{m_i!}. \tag{5.30}$$

Thus, from (5.28)–(5.30), we conclude that

$$\lim_{n\to\infty} P_n^{(\theta)}(C_1^{(n)} = m_1, C_2^{(n)} = m_2, \ldots, C_j^{(n)} = m_j) = \prod_{i=1}^{j} e^{-\frac{\theta}{i}} \frac{(\frac{\theta}{i})^{m_i}}{m_i!},$$

completing the proof of Theorem 5.2. □

Exercise 5.1. Verify that Proposition 5.1 follows from the definition of $P_n^{(\theta)}$ along with Proposition 5.2 and Lemma 5.1.

Exercise 5.2. Show that (5.4) is equivalent to

$$\lim_{n\to\infty} P_n^{(\theta)}(C_1^{(n)} \le m_1, C_2^{(n)} \le m_2, \ldots, C_j^{(n)} \le m_j) = \sum_{0\le r_1\le m_1,\ldots,0\le r_j\le m_j} \prod_{i=1}^{j} e^{-\frac{\theta}{i}} \frac{(\frac{\theta}{i})^{r_i}}{r_i!},$$

$$m_i \ge 0,\ i = 1, \ldots, j. \tag{5.31}$$

Exercise 5.3. In this exercise you will show directly that (5.5) follows from (5.4).

(a) Fix an integer $j \ge 2$. Use (5.31) to show that for any $\epsilon > 0$, there exists an N_ϵ such that if $n \ge N_\epsilon$ and $m \ge N_\epsilon$, then

$$P_n^{(\theta)}(C_i^{(n)} > m, \text{ for some } i \in [j]) < \epsilon. \tag{5.32}$$

(b) From (5.31) and (5.32), deduce that (5.31) also holds if some of the m_i are equal to ∞.

(c) Prove that (5.5) follows from (5.4).

Exercise 5.4. This exercise gives an alternative probabilistic proof of Proposition 5.3. A uniformly random ($\theta = 1$) permutation $\sigma \in S_n$ can be constructed in the following manner via its cycles. We begin with the number 1. Now we randomly choose a number from $[n]$. If we chose j, then we declare that $\sigma_1 = j$. This is the first stage of the construction. If $j \ne 1$, then we randomly choose a number from $[n] - \{j\}$. If we chose k, then we declare that $\sigma_j = k$. This is the second stage of the construction. If $k \ne 1$, then we randomly choose a number from $[n] - \{j, k\}$. We continue like this until we finally choose 1, which closes the cycle. For example, if after k we chose 1, then the permutation σ would contain the cycle $(1jk)$. Once we close a cycle, we begin again, starting with the smallest number that has not yet been used. We continue like this for n stages, at which point the permutation σ has been defined completely.

(a) The above construction has n stages. Show that the probability of completing a cycle on the jth stage is $\frac{1}{n+1-j}$. Thus, letting

$$X_j^{(n)} = \begin{cases} 1, & \text{if a cycle was completed at stage } j; \\ 0, & \text{otherwise,} \end{cases}$$

it follows that $X_j^{(n)} \sim \text{Ber}(\frac{1}{n+1-j})$.

(b) Argue that $\{X_j^{(n)}\}_{j=1}^n$ are independent.

(c) Show that the number of cycles $N^{(n)}$ can be represented as $N^{(n)} = \sum_{j=1}^n X_j^{(n)}$, thereby proving Proposition 5.3 in the case $\theta = 1$.

(d) Let $\theta \in (0, \infty)$. Amend the above construction as follows. At any stage j, close the cycle with probability $\frac{\theta}{n+\theta-j}$, and choose any other particular number that has not yet been used with probability $\frac{1}{n+\theta-j}$. Show that this construction yields a permutation distributed according to $P_n^{(\theta)}$, and use the above reasoning to prove Proposition 5.3 for all $\theta > 0$.

Exercise 5.5. (a) Show that if $\sum_{i=0}^\infty |b_i| < \infty$ and the triangular array $\{c_{n,i} : i = 0, 1, \ldots, n;\ n = 0, 1, \ldots\}$ is bounded and satisfies $\lim_{n\to\infty} c_{n,n-i} = 1$, for all i, then $\lim_{n\to\infty} \sum_{i=1}^n b_i c_{n,n-i} = \sum_{i=1}^\infty b_i$. Then use this to prove (5.21) in the case that $\theta > 1$.

(b) Show that if $\theta \in (0, 1)$, then $\frac{n!}{(n-i)!} \frac{\theta^{(n-i)}}{\theta^{(n)}} \le \frac{n}{n-i}$, if $i < n$. Also, $\frac{n!}{0!} \frac{\theta^{(0)}}{\theta^{(n)}} \le n$, where we recall, $\theta^{(0)} = 1$.

(c) Show that if $\sum_{i=0}^\infty |b_i| s^i < \infty$, where $s > 1$, then $|b_i| \le s^{-i}$, for all large i.

(d) For $\theta \in (0, 1)$, prove (5.21) as follows. Break the sum $\frac{n!}{\theta^{(n)}} \sum_{i=0}^n b_i \frac{\theta^{(n-i)}}{(n-i)!}$ into three parts—from $i = 0$ to $i = N$, from $i = N + 1$ to $i = [\frac{n}{2}]$, and from $i = [\frac{n}{2}] + 1$ to $i = n$. Use the reasoning in the proof of (a) to show that by choosing N sufficiently large, the limit as $n \to \infty$ of the first part can be made arbitrarily close to $\sum_{i=0}^\infty b_i$. Use the fact that $\sum_{i=0}^\infty |b_i| < \infty$ to show that by choosing N sufficiently large, the $\limsup_{n\to\infty}$ of the second part can be made arbitrarily small. Use (b) and (c) to show that the limit as $n \to \infty$ of the third part is 0.

Exercise 5.6. Prove that $\bar{b}_i \ge |b_i|$, where $\{b_i\}_{i=0}^\infty$ and $\{\bar{b}_i\}_{i=0}^\infty$ are defined in (5.23) and (5.24).

Exercise 5.7. Make a small change in the proof of Theorem 5.2 to show that (5.5) holds.

Exercise 5.8. Consider the uniform probability measure $P_n^{(1)}$ on S_n and let $E_n^{(1)}$ denote the expectation under $P_n^{(1)}$. Let $X_n = X_n(\sigma)$ be the random variable denoting the number of nearest neighbor pairs in the permutation $\sigma \in S_n$, and let $Y_n = Y_n(\sigma)$ be the random variable denoting the number of nearest neighbor triples in $\sigma \in S_n$. (A nearest neighbor pair for σ is a pair $k, k+1$, with $k \in [n-1]$, such that $\sigma_i = k$ and $\sigma_{i+1} = k+1$, for some $i \in [n-1]$, and a nearest neighbor triple is a triple $(k, k+1, k+2)$ with $k \in [n-2]$ such that $\sigma_i = k$, $\sigma_{i+1} = k+1$ and $\sigma_{i+2} = k+2$, for some $i \in [n-2]$.)

(a) Show that $E_n^{(1)} X_n = 1$, for all n. (Hint: Represent X_n as the sum of indicator random variables $\{I_k\}_{k=1}^{n-1}$, where $I_k(\sigma)$ is equal to 1 if $k, k+1$ is a nearest neighbor pair in σ and is equal to 0 otherwise.) It can be shown that the distribution of X_n converges weakly to the Pois(1) distribution as $n \to \infty$; see [17].
(b) Show that $\lim_{n\to\infty} E_n^{(1)} Y_n = 0$ and conclude that $\lim_{n\to\infty} P_n^{(1)}(Y_n = 0) = 1$.

Chapter Notes

In this chapter we investigated the limiting distribution as $n \to \infty$ of the random vector denoting the number of cycles of lengths 1 through j in a random permutation from S_n. It is very interesting and more challenging to investigate the limiting distribution of the random vector denoting the j longest cycles or, alternatively, the j shortest cycles. For everything you want to know about cycles in random permutations, and lots of references, see the book by Arratia et al. [6]. Our approach in this chapter was almost completely combinatorial, through the use of generating functions. Such methods are used occasionally in [6], but the emphasis is on more sophisticated probabilistic analysis. Our method is similar to the generating function approach of Wilf in [34], which deals only with the case $\theta = 1$. For an expository account of the intertwining of combinatorial objects with stochastic processes, see the lecture notes of Pitman [30].

Chapter 6
Chebyshev's Theorem on the Asymptotic Density of the Primes

Let $\pi(n)$ denote the number of primes that are no larger than n; that is,

$$\pi(n) = \sum_{p \le n} 1,$$

where *here and elsewhere in this chapter and the next two, the letter p in a summation denotes a prime.* Euclid proved that there are infinitely many primes: $\lim_{n\to\infty} \pi(n) = \infty$. The asymptotic density of the primes is 0; that is,

$$\lim_{n\to\infty} \frac{\pi(n)}{n} = 0.$$

The *prime number theorem* gives the leading order asymptotic behavior of $\pi(n)$. It states that

$$\lim_{n\to\infty} \frac{\pi(n)\log n}{n} = 1.$$

This landmark result was proved in 1896 independently by J. Hadamard and by C.J. de la Vallée Poussin. Their proofs used contour integration and Cauchy's theorem from analytic function theory. A so-called "elementary" proof, that is, a proof that does not use analytic function theory, was given by P. Erdős and A. Selberg in 1949. Although their proof uses only elementary methods, it is certainly more involved than the proofs of Hadamard and de la Vallée Poussin. We will not prove the prime number theorem in this book. In this chapter we prove a precursor of the prime number theorem, due to Chebyshev in 1850. Chebyshev was the first to prove that $\pi(n)$ grows on the order $\frac{n}{\log n}$. Chebyshev's methods were ingenious but entirely elementary. Given the truly elementary nature of his approach, it is quite impressive how close his result is to the prime number theorem. Here is Chebyshev's result.

R.G. Pinsky, *Problems from the Discrete to the Continuous*, Universitext,
DOI 10.1007/978-3-319-07965-3_6, © Springer International Publishing Switzerland 2014

Theorem 6.1 (Chebyshev).

$$0.693 \approx \log 2 \le \liminf_{n\to\infty} \frac{\pi(n)\log n}{n} \le \limsup_{n\to\infty} \frac{\pi(n)\log n}{n} \le \log 4 \approx 1.386.$$

Chebyshev's result is not the type of result we are emphasizing in this book, since it is not an exact asymptotic result but rather only an estimate. We have included the result because we will need it to prove Mertens' theorems in Chap. 7, and one of Mertens' theorems will be used to prove the Hardy–Ramanujan theorem in Chap. 8.

Define *Chebyshev's θ-function* by

$$\theta(n) = \sum_{p\le n} \log p. \tag{6.1}$$

Chebyshev realized that an understanding of the asymptotic behavior of $\theta(n)$ allows one to infer the asymptotic behavior of $\pi(n)$ (and vice versa), and that the direct asymptotic analysis of the function θ is much more tractable than that of the function π, because the sum of logarithms is the logarithm of the product. Indeed, note that

$$\theta(n) = \log \prod_{p\le n} p. \tag{6.2}$$

We will give an exceedingly simple proof of the following result, which links the asymptotic behavior of θ to that of π.

Proposition 6.1.

(i) $\liminf_{n\to\infty} \frac{\theta(n)}{n} = \liminf_{n\to\infty} \frac{\pi(n)\log n}{n}$;
(ii) $\limsup_{n\to\infty} \frac{\theta(n)}{n} = \limsup_{n\to\infty} \frac{\pi(n)\log n}{n}$.

Proof. We have the trivial inequality

$$\theta(n) = \sum_{p\le n} \log p \le \pi(n)\log n.$$

Dividing this by n and letting $n \to \infty$, we obtain

$$\liminf_{n\to\infty} \frac{\theta(n)}{n} \le \liminf_{n\to\infty} \frac{\pi(n)\log n}{n}; \qquad \limsup_{n\to\infty} \frac{\theta(n)}{n} \le \limsup_{n\to\infty} \frac{\pi(n)\log n}{n}. \tag{6.3}$$

We have for $\epsilon \in (0,1)$,

$$\theta(n) \ge \sum_{[n^{1-\epsilon}]<p\le n} \log p \ge \Big(\pi(n) - \pi([n^{1-\epsilon}])\Big)\log n^{1-\epsilon} \ge$$
$$(1-\epsilon)\Big(\pi(n)\log n - [n^{1-\epsilon}]\log n\Big),$$

where the last inequality comes from the trivial fact that $\pi(y) \le y$. Dividing this by n and letting $n \to \infty$, and using the fact that $\epsilon \in (0,1)$ is arbitrary, we obtain

$$\liminf_{n\to\infty} \frac{\theta(n)}{n} \ge \liminf_{n\to\infty} \frac{\pi(n)\log n}{n}; \qquad \limsup_{n\to\infty} \frac{\theta(n)}{n} \ge \limsup_{n\to\infty} \frac{\pi(n)\log n}{n}. \tag{6.4}$$

The proposition follows from (6.3) and (6.4). □

The following theorem gives an upper bound on the Chebyshev θ-function.

Theorem 6.2. $\frac{\theta(n)}{n} \le \log 4,\ n \ge 1.$

Proof. The proof is by induction, the inductive hypothesis being that $\theta(n) \le n \log 4$. Note that the hypothesis holds for $n = 1, 2$. If $n + 1 \ge 3$ is even, then $\theta(n+1) = \theta(n) \le n \log 4 \le (n+1)\log 4$, where the first inequality comes from the inductive hypothesis. If $n + 1$ is odd, then write $n + 1 = 2m + 1$, and note that the binomial coefficient $\binom{2m+1}{m} = \frac{(2m+1)(2m)\cdots(m+2)}{m!}$ is divisible by every prime between $m + 2$ and $2m + 1$ (since all such primes appear in the numerator of the latter expression, but not in the denominator). Since $\binom{2m+1}{m}$ is a positive integer (all binomial coefficients are integers) which contains as factors all the primes between $m + 2$ and $2m + 1$, we have

$$\prod_{m+2\le p\le 2m+1} p \le \binom{2m+1}{m}. \tag{6.5}$$

By the binomial formula,

$$2^{2m+1} = (1+1)^{2m+1} = \sum_{j=0}^{2m+1} \binom{2m+1}{j} \ge \binom{2m+1}{m} + \binom{2m+1}{m+1} = 2\binom{2m+1}{m};$$

thus,

$$\binom{2m+1}{m} \le 2^{2m}. \tag{6.6}$$

From (6.2), (6.5), and (6.6) we have

$$\theta(2m+1) - \theta(m+1) = \log \prod_{m+2\le p\le 2m+1} p \le \log 2^{2m} = m \log 4. \tag{6.7}$$

From (6.7) and the inductive hypothesis, we have

$$\theta(2m+1) \le \theta(m+1) + m\log 4 \le (m+1)\log 4 + m \log 4 = (2m+1)\log 4;$$

that is, $\theta(n+1) \le (n+1)\log 4$. □

As we noted above, the direct asymptotic analysis of θ is much more tractable than that of π, and Theorem 6.2 carried out an upper bound analysis for θ. It turns out that for the lower-bound analysis it is better to work with *Chebyshev's ψ-function* instead of Chebyshev's θ-function. One defines

$$\psi(n) = \sum_{p^k \le n, k \ge 1} \log p. \tag{6.8}$$

That is, in the sum above, a term $\log p$ appears for every prime p and integer $k \ge 1$ for which $p^k \le n$. So, for example, $\psi(14) = 3\log 2 + 2\log 3 + \log 5 + \log 7 + \log 11 + \log 13$. Of course, $\psi(n) \ge \theta(n)$. We show now that θ and ψ have the same asymptotic behavior.

Proposition 6.2.

(i) $\liminf_{n\to\infty} \frac{\theta(n)}{n} = \liminf_{n\to\infty} \frac{\psi(n)}{n}$;
(ii) $\limsup_{n\to\infty} \frac{\theta(n)}{n} = \limsup_{n\to\infty} \frac{\psi(n)}{n}$.

Proof. Since $\psi(n) \ge \theta(n)$, we have

$$\liminf_{n\to\infty} \frac{\theta(n)}{n} \le \liminf_{n\to\infty} \frac{\psi(n)}{n}; \qquad \limsup_{n\to\infty} \frac{\theta(n)}{n} \le \limsup_{n\to\infty} \frac{\psi(n)}{n}. \tag{6.9}$$

Since $2^k \le n$ if and only if $k\log 2 \le \log n$, or equivalently, $k \le [\frac{\log n}{\log 2}]$, it follows that $p^k > n$ for every prime p and every $k > [\frac{\log n}{\log 2}]$; thus

$$\psi(n) - \theta(n) = \sum_{p^k \le n, k \ge 2} \log p = \sum_{k=2}^{[\frac{\log n}{\log 2}]} \sum_{p \le n^{\frac{1}{k}}} \log p =$$

$$\sum_{k=2}^{[\frac{\log n}{\log 2}]} \theta([n^{\frac{1}{k}}]) \le \frac{\log n}{\log 2}\theta([n^{\frac{1}{2}}]). \tag{6.10}$$

Now trivially, $\theta(k) = \sum_{p \le k} \log p \le k \log k$. Using this with (6.10) gives

$$\psi(n) - \theta(n) \le \frac{(\log n)^2 n^{\frac{1}{2}}}{2\log 2}. \tag{6.11}$$

From (6.11) it follows that

$$\liminf_{n\to\infty} \frac{\theta(n)}{n} \ge \liminf_{n\to\infty} \frac{\psi(n)}{n}; \qquad \limsup_{n\to\infty} \frac{\theta(n)}{n} \ge \limsup_{n\to\infty} \frac{\psi(n)}{n}. \tag{6.12}$$

The proposition follows from (6.9) and (6.12). □

Remark. The bound obtained in (6.11) can be improved by replacing the trivial bound on θ, namely, $\theta(k) \leq k \log k$, by the bound obtained from Theorem 6.2.

We will carry out a lower-bound analysis of ψ. This will be somewhat more involved than the upper bound analysis for θ but still entirely elementary. For $n \in \mathbb{N}$ and p a prime, let $v_p(n)$ denote the largest exponent k such that $p^k|n$. One calls $v_p(n)$ the *p-adic value of n*. It follows from the definition of v_p that any positive integer n can be written as

$$n = \prod_p p^{v_p(n)}. \tag{6.13}$$

In Exercise 6.1 the reader is asked to prove the following simple formula:

$$v_p(mn) = v_p(m) + v_p(n), \ m, n \in \mathbb{N}. \tag{6.14}$$

From (6.14) it follows that

$$v_p(n!) = \sum_{m=1}^{n} v_p(m). \tag{6.15}$$

We will need the following result.

Proposition 6.3.

$$v_p(n!) = \sum_{k=1}^{\infty} [\frac{n}{p^k}].$$

Proof. We can write

$$v_p(m) = \sum_{1 \leq k < \infty, p^k|m} 1.$$

Using this with (6.15), we have

$$v_p(n!) = \sum_{m=1}^{n} v_p(m) = \sum_{m=1}^{n} \sum_{1 \leq k < \infty, p^k|m} 1 = \sum_{k=1}^{\infty} \sum_{1 \leq m \leq n, p^k|m} 1. \tag{6.16}$$

If $p^k > n$, then obviously there is no $m \in [n]$ for which $p^k|m$. If $p^k \leq n$, then the integers $m \in [n]$ for which $p^k|m$ are the $[\frac{n}{p^k}]$ integers $p^k, \ldots, [\frac{n}{p^k}]p^k$. Thus, $\sum_{1 \leq m \leq n, p^k|m} 1 = [\frac{n}{p^k}]$. Substituting this in (6.16) completes the proof of the proposition. □

We can now carry out a lower-bound analysis of ψ.

Theorem 6.3.

$$\liminf_{n\to\infty}\frac{\psi(n)}{n}\ge\log 2.$$

Proof. Consider the binomial coefficient $\binom{2n}{n}=\frac{(2n)!}{(n!)^2}$. Using (6.13) we have

$$\binom{2n}{n}=\frac{(2n)!}{(n!)^2}=\prod_p p^{v_p((2n)!)-2v_p(n!)}=\prod_{p\le 2n}p^{v_p((2n)!)-2v_p(n!)},\tag{6.17}$$

where the final equality comes from the fact that neither $(2n)!$ nor $n!$ has a prime factor larger than $2n$. From Proposition 6.3, we have

$$v_p((2n)!)-2v_p(n!)=\sum_{k=1}^{\infty}\Big([\frac{2n}{p^k}]-2[\frac{n}{p^k}]\Big).\tag{6.18}$$

Of course, $[\frac{2n}{p^k}]=[\frac{n}{p^k}]=0$ if $p^k>2n$, that is, if $k>[\frac{\log 2n}{\log p}]$. Thus, in the summation over k above, we may replace the upper limit ∞ by $[\frac{\log 2n}{\log p}]$. Furthermore, it is easy to verify that $[2x]-2[x]$ is equal to either 0 or 1, for all real numbers x. From these two facts we obtain from (6.18) the estimate

$$0\le v_p((2n)!)-2v_p(n!)\le[\frac{\log 2n}{\log p}].\tag{6.19}$$

From (6.17) and (6.19) we have the estimate

$$\binom{2n}{n}\le\prod_{p\le 2n}p^{[\frac{\log 2n}{\log p}]}.\tag{6.20}$$

On the other hand we have the easy estimate

$$\binom{2n}{n}\ge\frac{2^{2n}}{2n}.\tag{6.21}$$

To prove (6.21), note that the middle binomial coefficient $\binom{2n}{n}$ maximizes $\binom{2n}{k}$ over $k\in[2n]$. The reader is asked to prove this in Exercise 6.2. Thus, we have

$$2^{2n}=(1+1)^{2n}=\sum_{k=0}^{2n}\binom{2n}{k}=2+\sum_{k=1}^{2n-1}\binom{2n}{k}\le 2+(2n-1)\binom{2n}{n}\le 2n\binom{2n}{n}.$$

From (6.20) and (6.21), we conclude that

$$\frac{2^{2n}}{2n} \leq \prod_{p \leq 2n} p^{[\frac{\log 2n}{\log p}]}$$

or, equivalently,

$$2n \log 2 - \log 2n \leq \sum_{p \leq 2n} [\frac{\log 2n}{\log p}] \log p. \tag{6.22}$$

Recalling from (6.8) that $\psi(2n) = \sum_{p^k \leq 2n, k \geq 1} \log p$, it follows that the summand $\log p$ appears in $\psi(2n)$ one time for each $k \geq 1$ that satisfies $p^k \leq 2n$; that is, the summand $\log p$ appears $[\frac{\log 2n}{\log p}]$ times. Thus, the right hand side of (6.22) is equal to $\psi(2n)$, giving the inequality

$$\psi(2n) \geq 2n \log 2 - \log 2n. \tag{6.23}$$

Of course then we also have

$$\psi(2n+1) \geq 2n \log 2 - \log 2n. \tag{6.24}$$

Dividing (6.23) by $2n$ and dividing (6.24) by $2n+1$, and letting $n \to \infty$, we conclude that

$$\liminf_{n \to \infty} \frac{\psi(n)}{n} \geq \log 2,$$

which completes the proof of the theorem. □

We can now prove Chebyshev's theorem in one line.

Proof of Theorem 6.1. The upper bound follows from Theorem 6.2 and part (ii) of Proposition 6.1, while the lower bound follows from Theorem 6.3, part (i) of Proposition 6.2, and part (i) of Proposition 6.1. □

Exercise 6.1. Prove (6.14): $v_p(mn) = v_p(m) + v_p(n)$, $m, n \in \mathbb{N}$.

Exercise 6.2. Prove that $\binom{2n}{n} = \max_{k \in [2n]} \binom{2n}{k}$.

Exercise 6.3. *Bertrand's postulate* states that for each positive integer n, there exists a prime in the interval $(n, 2n)$. This result was first proven by Chebyshev. Use the upper and lower bounds obtained in this chapter for Chebyshev's θ-function to prove the following weak form of Bertrand's postulate: For every $\epsilon > 0$, there exists an $n_0(\epsilon)$ such that for every $n \geq n_0(\epsilon)$ there exists a prime in the interval $(n, (2+\epsilon)n)$.

Chapter Notes

Chebyshev also proved that if $\lim_{n\to\infty} \frac{\pi(n)\log n}{n}$ exists, then this limit must be equal to 1. For a proof, see Tenenbaum's book [33]. Late in his life, in a letter, Gauss recollected that in the early 1790s, when he was 15 or 16, he conjectured the prime number theorem; however, he never published the conjecture. The theorem was conjectured by Dirichlet in 1838. For some references for further reading, see the notes at the end of Chap. 8.

Chapter 7
Mertens' Theorems on the Asymptotic Behavior of the Primes

Given a sequence of positive numbers $\{a_n\}_{n=1}^{\infty}$ satisfying $\lim_{n\to\infty} a_n = \infty$, one way to measure the rate at which the sequence approaches ∞ is to consider the rate at which the series $\sum_{j=1}^{n} \frac{1}{a_j}$ grows. For $a_j = j$, it is well known that the *harmonic series* $\sum_{j=1}^{n} \frac{1}{j}$ satisfies $\sum_{j=1}^{n} \frac{1}{j} = \log n + O(1)$ as $n \to \infty$. How does the harmonic series of the primes behave? The goal of this chapter is to prove a theorem known as *Mertens' second theorem.*

Theorem 7.1.

$$\sum_{p\leq n} \frac{1}{p} = \log\log n + O(1), \text{ as } n \to \infty.$$

Mertens' second theorem will play a key role in the proof of the Hardy–Ramanujan theorem in Chap. 8. For our proof of Mertens' second theorem, we will need a result known as *Mertens' first theorem.*

Theorem 7.2.

$$\sum_{p\leq n} \frac{\log p}{p} = \log n + O(1), \text{ as } n \to \infty.$$

We now prove Mertens' two theorems.

Proof of Mertens' first theorem. We will analyze the asymptotic behavior of $\log n!$ in two different ways. Comparing the two results will prove the theorem. First we show that

$$\log n! = n \log n + O(n), \text{ as } n \to \infty. \tag{7.1}$$

R.G. Pinsky, *Problems from the Discrete to the Continuous*, Universitext,
DOI 10.1007/978-3-319-07965-3_7, © Springer International Publishing Switzerland 2014

We note that (7.1) follows from Stirling's formula: $n! \sim n^n e^{-n}\sqrt{2\pi n}$. However, we certainly don't need such a precise estimate of $n!$ to obtain (7.1). We give a quick direct proof of (7.1). Consider an integer $m \geq 2$ and $x \in [m-1, m]$. Integrating the inequality $\log(m-1) \leq \log x \leq \log m$ over $x \in [m-1, m]$ gives

$$\log(m-1) \leq \int_{m-1}^{m} \log x \, dx \leq \log m,$$

which we rewrite as

$$0 \leq \log m - \int_{m-1}^{m} \log x \, dx \leq \log m - \log(m-1).$$

Summing this inequality from $m = 2$ to $m = n$, and noting that the resulting series on the right hand side is telescopic, we obtain

$$0 \leq \log n! - \int_{1}^{n} \log x \, dx \leq \log n. \tag{7.2}$$

An integration by parts shows that $\int_1^n \log x \, dx = n \log n - n + 1$. Substituting this in (7.2) gives

$$n \log n - n + 1 \leq \log n! \leq n \log n - n + 1 + \log n,$$

which completes the proof of (7.1).

To analyze $\log n!$ in another way, we utilize the function $v_p(n)$ introduced in Chap. 6. Recall that $v_p(n)$, the p-adic value of n, is equal to the largest exponent k such that $p^k | n$ and that by the definition of v_p, we have $n = \prod_p p^{v_p(n)} = \prod_{p \leq m} p^{v_p(n)}$, for any integer m that is greater than or equal to the largest prime divisor of n. Recall that Proposition 6.3 states that

$$v_p(n!) = \sum_{k=1}^{\infty} [\frac{n}{p^k}].$$

Thus, we have

$$n! = \prod_{p \leq n} p^{v_p(n!)} = \prod_{p \leq n} p^{\sum_{k=1}^{\infty} [\frac{n}{p^k}]},$$

and

$$\log n! = \sum_{p \leq n} (\sum_{k=1}^{\infty} [\frac{n}{p^k}]) \log p = \sum_{p \leq n} [\frac{n}{p}] \log p + \sum_{p \leq n} (\sum_{k=2}^{\infty} [\frac{n}{p^k}]) \log p. \tag{7.3}$$

We now analyze the two terms on the right hand of (7.3), beginning with the second term. We have

$$\sum_{k=2}^{\infty}[\frac{n}{p^k}] \leq n\sum_{k=2}^{\infty}\frac{1}{p^k} = n\frac{\frac{1}{p^2}}{1-\frac{1}{p}} = \frac{n}{p(p-1)}.$$

Thus, we obtain

$$\sum_{p\leq n}(\sum_{k=2}^{\infty}[\frac{n}{p^k}])\log p \leq n\sum_{p\leq n}\frac{\log p}{p(p-1)} \leq Cn, \tag{7.4}$$

for some constant $C > 0$, the latter inequality following from the fact that $\sum_{p\leq n}\frac{\log p}{p(p-1)} < \sum_{m=2}^{\infty}\frac{\log m}{m(m-1)} < \infty$. We write the first term on the right hand side of (7.3) as

$$\sum_{p\leq n}[\frac{n}{p}]\log p = n\sum_{p\leq n}\frac{\log p}{p} - \sum_{p\leq n}(\frac{n}{p} - [\frac{n}{p}])\log p. \tag{7.5}$$

Recalling that Theorem 6.2 gives $\theta(n) \leq (\log 4)n$, we can estimate the second term on the right hand side of (7.5) by

$$0 \leq \sum_{p\leq n}(\frac{n}{p} - [\frac{n}{p}])\log p \leq \sum_{p\leq n}\log p = \theta(n) \leq (\log 4)n. \tag{7.6}$$

From (7.3)–(7.6), we conclude that

$$\log n! = n\sum_{p\leq n}\frac{\log p}{p} + O(n), \text{ as } n \to \infty. \tag{7.7}$$

Comparing (7.1) with (7.7) allows us to conclude that $\sum_{p\leq n}\frac{\log p}{p} = \log n + O(1)$, completing the proof of Mertens' first theorem. □

In order to use Mertens' first theorem to prove his second theorem, we need to introduce *Abel summation*, a tool that is used extensively in number theory. Abel summation is a discrete version of integration by parts. It appears in a variety of guises, the following of which is the most suitable in the present context.

Proposition 7.1 (Abel Summation). *Let $j_0, n \in \mathbb{Z}$ with $j_0 < n$. Let $a : [j_0, n] \cap \mathbb{Z} \to \mathbb{R}$, and let $A : [j_0, n] \to \mathbb{R}$ be defined by $A(t) = \sum_{k=j_0}^{[t]} a(k)$. Let $f : [j_0, n] \to \mathbb{R}$ be continuously differentiable. Then*

$$\sum_{j_0<r\leq n} a(r)f(r) = A(n)f(n) - A(j_0)f(j_0) - \int_{j_0}^{n} A(t)f'(t)\,dt. \tag{7.8}$$

Remark. Since $A(j_0) = a(j_0)$, we could also write the above formula in the more compact form

$$\sum_{j_0 \leq r \leq n} a(r)f(r) = A(n)f(n) - \int_{j_0}^{n} A(t)f'(t)\,dt. \tag{7.9}$$

The form in the proposition of course mimics the standard integration by parts formula.

Proof. Since A is constant between integers, we have

$$\int_{j_0}^{n} A(t)f'(t)\,dt = \sum_{r=j_0}^{n-1} \int_{r}^{r+1} A(t)f'(t)\,dt = \sum_{r=j_0}^{n-1} A(r)\big(f(r+1) - f(r)\big). \tag{7.10}$$

Substituting for A in the last term on the right hand side, and interchanging the order of the resulting summation, we obtain

$$\sum_{r=j_0}^{n-1} A(r)\big(f(r+1) - f(r)\big) = \sum_{r=j_0}^{n-1} \Big(\sum_{k=j_0}^{r} a(k)\Big)\big(f(r+1) - f(r)\big) =$$

$$\sum_{k=j_0}^{n-1} a(k) \sum_{r=k}^{n-1} \big(f(r+1) - f(r)\big) = \sum_{k=j_0}^{n-1} a(k)\big(f(n) - f(k)\big) =$$

$$A(n-1)f(n) - \sum_{k=j_0}^{n-1} a(k)f(k). \tag{7.11}$$

From (7.10) and (7.11) we obtain

$$\int_{j_0}^{n} A(t)f'(t)\,dt = A(n-1)f(n) - \sum_{k=j_0}^{n-1} a(k)f(k).$$

Substituting this in the right hand side of (7.8) gives

$$A(n)f(n) - A(j_0)f(j_0) - \int_{j_0}^{n} A(t)f'(t)\,dt =$$

$$A(n)f(n) - A(j_0)f(j_0) - A(n-1)f(n) + \sum_{k=j_0}^{n-1} a(k)f(k) = \sum_{k=j_0+1}^{n} a(k)f(k), \tag{7.12}$$

which proves the proposition. □

Proof of Mertens' second theorem. Let

$$a(n) = \begin{cases} \frac{\log p}{p}, \text{ if } n = p; \\ 0, \text{ otherwise,} \end{cases}$$

and let

$$f(t) = \frac{1}{\log t}, \ t > 1.$$

We use Abel summation in the form (7.9) with $j_0 = 2$. By Mertens' first theorem, we have

$$A(t) = \sum_{k=2}^{[t]} a(k) = \sum_{p \le [t]} \frac{\log p}{p} = \log t + O(1), \text{ as } t \to \infty. \tag{7.13}$$

Thus, we obtain from (7.9) and (7.13),

$$\sum_{p \le n} \frac{1}{p} = \sum_{p \le n} \frac{\log p}{p} \frac{1}{\log p} = \sum_{2 \le r \le n} a(r) f(r) = A(n) f(n) - \int_2^n A(t) f'(t)\, dt =$$

$$\frac{\log n + O(1)}{\log n} + \int_2^n \frac{\log t + O(1)}{t (\log t)^2}\, dt. \tag{7.14}$$

We have

$$\int_2^n \frac{1}{t \log t}\, dt = \log\log t\,|_2^n = \log\log n - \log\log 2,$$

and since $\int \frac{1}{t(\log t)^2}\, dt = -\frac{1}{\log t}$, we have

$$\int_2^\infty \frac{1}{t(\log t)^2}\, dt < \infty.$$

Using these two facts in (7.14) gives

$$\sum_{p \le n} \frac{1}{p} = \log\log n + O(1), \text{ as } n \to \infty, \tag{7.15}$$

completing the proof of Mertens' second theorem. □

Exercise 7.1. (a) Use Mertens' first theorem and Abel summation to prove that

$$\sum_{p\leq n} \frac{\log^2 p}{p} = \frac{1}{2}\log^2 n + O(\log n).$$

(Hint: Write $\sum_{p\leq n} \frac{\log^2 p}{p} = \sum_{1\leq r\leq n} a(r)\log r$, where $a(r)$ is as in the proof of Mertens' second theorem.)

(b) Use induction and the result in (a) to prove that

$$\sum_{p\leq n} \frac{\log^k p}{p} = \frac{1}{k}\log^k n + O(\log^{k-1} n),$$

for all positive integers k.

Exercise 7.2. Proposition 6.1 in Chap. 6 showed that the two statements, $\sum_{p\leq n} \log p \sim n$ and $\pi(n) = \sum_{p\leq n} 1 \sim \frac{n}{\log n}$, can easily be derived one from the other. The prime number theorem cannot be derived from Mertens' second theorem. Derive Mertens' second theorem in the form $\sum_{p\leq n} \frac{1}{p} \sim \log\log n$ from the prime number theorem, $\pi(n) \sim \frac{n}{\log n}$. (Hint: Use Abel summation.)

Chapter Notes

The two theorems in this chapter were proven by F. Mertens in 1874. For some references for further reading, see the notes at the end of Chap. 8.

Chapter 8
The Hardy–Ramanujan Theorem on the Number of Distinct Prime Divisors

Let $\omega(n)$ denote the number of distinct prime divisors of n; that is,

$$\omega(n) = \sum_{p|n} 1.$$

Thus, for example, $\omega(1) = 0$, $\omega(2) = 1$, $\omega(9) = 1$, $\omega(60) = 3$. The values of $\omega(n)$ obviously fluctuate wildly as $n \to \infty$, since $\omega(p) = 1$, for every prime p. However, there are not very many prime numbers, in the sense that the asymptotic density of the primes is 0. In this chapter we prove the Hardy–Ramanujan theorem, which in colloquial language states that "almost every" integer n has "approximately" $\log\log n$ distinct prime divisors. The meaning of "almost every" is that the asymptotic density of those integers n for which the number of distinct prime divisors is not "approximately" $\log\log n$ is zero. The meaning of "approximately" is that the actual number of distinct prime divisors of n falls in the interval $[\log\log n - (\log\log n)^{\frac{1}{2}+\delta}, \log\log n + (\log\log n)^{\frac{1}{2}+\delta}]$, where $\delta > 0$ is arbitrarily small.

Theorem 8.1 (Hardy–Ramanujan). *For every* $\delta > 0$,

$$\lim_{N\to\infty} \frac{|\{n \in [N] : |\omega(n) - \log\log n| \le (\log\log n)^{\frac{1}{2}+\delta}\}|}{N} = 1. \tag{8.1}$$

Remark. From the proof of the theorem, it is very easy to infer that the statement of the theorem is equivalent to the following statement: For every $\delta > 0$,

$$\lim_{N\to\infty} \frac{|\{n \in [N] : |\omega(n) - \log\log N| \le (\log\log N)^{\frac{1}{2}+\delta}\}|}{N} = 1.$$

R.G. Pinsky, *Problems from the Discrete to the Continuous*, Universitext,
DOI 10.1007/978-3-319-07965-3_8, © Springer International Publishing Switzerland 2014

While the statement of the theorem is probably more aesthetically pleasing than this latter statement, the latter statement is more practical. Thus, for example, take $\delta = .1$. Then for sufficiently large n, a very high percentage of the positive integers up to the astronomical number $N = e^{e^n}$ will have between $n - n^{.6}$ and $n + n^{.6}$ distinct prime factors. Let $n = 10^9$. We leave it to the interested reader to estimate the $O(1)$ terms appearing in the proofs of Mertens' theorems, and to keep track of how they appear in the proof of the Hardy–Ramanujan theorem below, and to conclude that over ninety percent of the positive integers up to $N = e^{e^{10^9}}$ have between $10^9 - (10^9)^{.6}$ and $10^9 + (10^9)^{.6}$ distinct prime factors. That is, over ninety percent of the positive integers up to $e^{e^{10^9}}$ have between $10^9 - 251,188$ and $10^9 + 251,188$ distinct prime factors.

Our proof of the Hardy–Ramanujan theorem will have a probabilistic flavor. For any positive integer N, let P_N denote the uniform probability measure on $[N]$; that is, $P_N(\{j\}) = \frac{1}{N}$, for $j \in [N]$. Then we may think of the distinct prime divisor function $\omega = \omega(n)$ as a random variable on the space $[N]$ with the probability measure P_N. For the sequel, note that when we write $P_N(\omega \in A)$, where $A \subset [N]$, what we mean is

$$P_N(\omega \in A) = P_N(\{n \in [N] : \omega(n) \in A\}) = \frac{|\{n \in [N] : \omega(n) \in A\}|}{N}.$$

Let E_N denote the expected value with respect to the measure P_N. The expected value of ω is given by

$$E_N\, \omega = \frac{1}{N} \sum_{n=1}^{N} \omega(n). \tag{8.2}$$

The second moment of ω is given by

$$E_N\, \omega^2 = \frac{1}{N} \sum_{n=1}^{N} \omega^2(n). \tag{8.3}$$

The variance $\mathrm{Var}_N(\omega)$ of ω is defined by

$$\mathrm{Var}_N(\omega) = E_N(\omega - E_N\, \omega)^2 = E_N\, \omega^2 - (E_N\, \omega)^2. \tag{8.4}$$

We will prove the Hardy–Ramanujan theorem by applying Chebyshev's inequality to the random variable ω:

$$P_N(|\omega - E_N\, \omega| \ge \lambda) \le \frac{\mathrm{Var}_N(\omega)}{\lambda^2}, \text{ for } \lambda > 0. \tag{8.5}$$

In order to implement this, we need to calculate $E_N\omega$ and $\mathrm{Var}_N(\omega)$ or, equivalently, $E_N\,\omega$ and $E_N\,\omega^2$. The next two theorems give the asymptotic behavior as $N \to \infty$ of $E_N\,\omega$ and of $E_N\,\omega^2$. The proofs of these two theorems will use Mertens' second theorem.

Theorem 8.2.

$$E_N\,\omega = \log\log N + O(1), \text{ as } N \to \infty.$$

Remark. Recall the definition of the average order of an arithmetic function, given in the remark following the number-theoretic proof of Theorem 2.1. Theorem 8.2 shows that the average order of ω, the function counting the number of distinct prime divisors, is given by the function $\log\log n$.

Proof. From the definition of the divisor function we have

$$\sum_{n=1}^{N}\omega(n) = \sum_{n=1}^{N}\sum_{p|n}1 = \sum_{p\leq N}\sum_{p|n,n\leq N}1 = \sum_{p\leq N}[\frac{N}{p}] =$$
$$N\sum_{p\leq N}\frac{1}{p} - \sum_{p\leq N}(\frac{N}{p} - [\frac{N}{p}]). \tag{8.6}$$

The second term above satisfies the inequality

$$0 \leq \sum_{p\leq N}(\frac{N}{p} - [\frac{N}{p}]) \leq \sum_{p\leq N}1 = \pi(N) \leq N. \tag{8.7}$$

(We could use Chebyshev's theorem (Theorem 6.1) to get the better bound $O(\frac{N}{\log N})$ on the right hand side above, but that wouldn't improve the order of the final bound we obtain for $E_N\omega$.) Mertens' second theorem (Theorem 7.1) gives

$$\sum_{p\leq N}\frac{1}{p} = \log\log N + O(1), \text{ as } N \to \infty. \tag{8.8}$$

From (8.6)–(8.8), we obtain

$$\sum_{n=1}^{N}\omega(n) = N\log\log N + O(N), \text{ as } N \to \infty,$$

and dividing this by N gives

$$E_N\,\omega = \log\log N + O(1), \text{ as } N \to \infty, \tag{8.9}$$

completing the proof of the theorem. □

Theorem 8.3.

$$E_N\,\omega^2 = (\log\log N)^2 + O(\log\log N), \text{ as } N \to \infty.$$

Remark. To prove the Hardy–Ramanujan theorem, we only need the upper bound

$$E_N\,\omega^2 \le (\log\log N)^2 + O(\log\log N), \text{ as } N \to \infty. \tag{8.10}$$

Proof. We have

$$\omega^2(n) = (\sum_{p|n} 1)^2 = (\sum_{p_1|n} 1)(\sum_{p_2|n} 1) = \sum_{\substack{p_1p_2|n \\ p_1\ne p_2}} 1 + \sum_{p|n} 1 = \sum_{\substack{p_1p_2|n \\ p_1\ne p_2}} 1 + \omega(n). \tag{8.11}$$

Thus,

$$\sum_{n=1}^{N} \omega^2(n) = \sum_{n=1}^{N} \sum_{\substack{p_1p_2|n \\ p_1\ne p_2}} 1 + \sum_{n=1}^{N} \omega(n). \tag{8.12}$$

The second term on the right hand side of (8.12) can be estimated by Theorem 8.2, giving

$$\sum_{n=1}^{N} \omega(n) = NE_N\,\omega = N\log\log N + O(N), \text{ as } N \to \infty. \tag{8.13}$$

To estimate the first term on the right hand side of (8.12), we write

$$\sum_{n=1}^{N} \sum_{\substack{p_1p_2|n \\ p_1\ne p_2}} 1 = \sum_{\substack{p_1p_2\le N \\ p_1\ne p_2}} \sum_{\substack{n\le N \\ p_1p_2|n}} 1 = \sum_{\substack{p_1p_2\le N \\ p_1\ne p_2}} [\frac{N}{p_1p_2}] =$$

$$N \sum_{\substack{p_1p_2\le N \\ p_1\ne p_2}} \frac{1}{p_1p_2} - \sum_{\substack{p_1p_2\le N \\ p_1\ne p_2}} \Big(\frac{N}{p_1p_2} - [\frac{N}{p_1p_2}]\Big). \tag{8.14}$$

The number of ordered pairs of distinct primes (p_1, p_2) such that $p_1p_2 \le N$ is of course equal to twice the number of such unordered pairs $\{p_1, p_2\}$. The *fundamental theorem of arithmetic* states that each integer has a unique factorization into primes; thus, if $p_1p_2 = p_3p_4$, then necessarily $\{p_1, p_2\} = \{p_3, p_4\}$. Consequently the number of unordered pairs $\{p_1, p_2\}$ such that $p_1p_2 \le N$ is certainly no greater than N. Thus, the second term on the right hand side of (8.14) satisfies

$$0 \le \sum_{\substack{p_1p_2\le N \\ p_1\ne p_2}} \Big(\frac{N}{p_1p_2} - [\frac{N}{p_1p_2}]\Big) \le \sum_{\substack{p_1p_2\le N \\ p_1\ne p_2}} 1 \le 2N. \tag{8.15}$$

Using Mertens' second theorem for the second inequality below, we bound from above the summation in the first term on the right hand side of (8.14) by

$$\sum_{\substack{p_1 p_2 \le N \\ p_1 \ne p_2}} \frac{1}{p_1 p_2} \le (\sum_{p \le N} \frac{1}{p})^2 \le \big(\log\log N + O(1)\big)^2, \text{ as } N \to \infty. \tag{8.16}$$

From (8.12)–(8.16), we conclude that (8.10) holds.

To complete the proof of the theorem, we need to show (8.10) with the reverse inequality. The easiest way to do this is to note simply that the variance is a nonnegative quantity. Thus,

$$E_N \omega^2 \ge (E_N \omega)^2 = \big(\log\log N + O(1)\big)^2 = (\log\log N)^2 + O(\log\log N),$$

where the first equality follows from Theorem 8.2. For an alternative proof, see Exercise 8.1. □

We now use Chebyshev's inequality along with the estimates in Theorems 8.2 and 8.3 to prove the Hardy–Ramanujan theorem.

Proof of Theorem 8.1. From Theorems 8.2 and 8.3 we have

$$\text{Var}_N(\omega) = E_N\, \omega^2 - (E_N \omega)^2 = (\log\log N)^2 + O(\log\log N) - \big(\log\log N + O(1)\big)^2 = O(\log\log N), \text{ as } N \to \infty. \tag{8.17}$$

Theorem 8.2 gives

$$E_N\, \omega = \log\log N + R_N, \text{ where } R_N \text{ is bounded as } N \to \infty. \tag{8.18}$$

Applying Chebyshev's inequality with $\lambda = (\log\log N)^{\frac{1}{2}+\delta}$, where $\delta > 0$, we obtain from (8.5), (8.17), and (8.18)

$$P_N\Big(|\omega - \log\log N - R_N| \ge (\log\log N)^{\frac{1}{2}+\delta}\Big) \le \frac{O(\log\log N)}{(\log\log N)^{1+2\delta}}, \text{ as } N \to \infty.$$

Thus,

$$\lim_{N\to\infty} P_N\Big(|\omega - \log\log N - R_N| \le (\log\log N)^{\frac{1}{2}+\delta}\Big) = 1. \tag{8.19}$$

Translating (8.19) back to the notation in the statement of the theorem, we have for every $\delta > 0$

$$\lim_{N\to\infty} \frac{|\{n \in [N] : |\omega(n) - \log\log N - R_N| \le (\log\log N)^{\frac{1}{2}+\delta}\}|}{N} = 1. \tag{8.20}$$

The main difference between (8.20) and the statement of the Hardy–Ramanujan theorem is that $\log\log N$ appears in (8.20) and $\log\log n$ appears in (8.1). Because $\log\log x$ is such a slowly varying function, this difference is not very significant. The remainder of the proof consists of showing that if (8.20) holds for all $\delta > 0$, then (8.1) also holds for all $\delta > 0$.

Fix an arbitrary $\delta > 0$. Using the fact that (8.20) holds with δ replaced by $\frac{\delta}{2}$, we will show that (8.1) holds for δ. This will then complete the proof of the theorem. The term R_N in (8.20) may vary with N, but it is bounded in absolute value, say by M. For $N^{\frac{1}{2}} \le n \le N$, we have

$$\log\log N - \log\log n \le \log\log N - \log\log N^{\frac{1}{2}} = \log 2. \tag{8.21}$$

Therefore, writing $\omega(n) - \log\log n = (\omega(n) - \log\log N - R_N) + (\log\log N - \log\log n) + R_N$, the triangle inequality and (8.21) give

$$|\omega(n)-\log\log n| \le |\omega(n)-\log\log N - R_N| + \log 2 + M, \text{ for } N^{\frac{1}{2}} \le n \le N. \tag{8.22}$$

Using (8.20) with δ replaced by $\frac{\delta}{2}$, along with (8.22) and the fact that $\lim_{N\to\infty} \frac{N^{\frac{1}{2}}}{N} = 0$, we have

$$\lim_{N\to\infty} \frac{|\{n \in [N] : |\omega(n) - \log\log n| \le (\log\log N)^{\frac{1}{2}+\frac{1}{2}\delta} + \log 2 + M\}|}{N} = 1. \tag{8.23}$$

By (8.21), it follows that $(\log\log n)^{\frac{1}{2}+\delta} \ge (\log\log N - \log 2)^{\frac{1}{2}+\delta}$, for $N^{\frac{1}{2}} \le n \le N$. Clearly, we have

$$(\log\log N - \log 2)^{\frac{1}{2}+\delta} \ge (\log\log N)^{\frac{1}{2}+\frac{1}{2}\delta} + \log 2 + M, \text{ for sufficiently large } N.$$

Thus,

$$(\log\log n)^{\frac{1}{2}+\delta} \ge (\log\log N)^{\frac{1}{2}+\frac{1}{2}\delta} + \log 2 + M, \text{ for } N^{\frac{1}{2}} \le n \le N \text{ and sufficiently large } N. \tag{8.24}$$

From (8.23), (8.24), and the fact that $\lim_{N\to\infty} \frac{N^{\frac{1}{2}}}{N} = 0$, we conclude that

$$\lim_{N\to\infty} \frac{|\{n \in [N] : |\omega(n) - \log\log n| \le (\log\log n)^{\frac{1}{2}+\delta}\}|}{N} = 1.$$

□

Exercise 8.1. Prove the lower bound

$$E_N\omega^2 \ge (\log\log N)^2 + O(\log\log N)$$

by using (8.12)–(8.15) and an inequality that begins with $\sum_{\substack{p_1 p_2 \le N \\ p_1 \ne p_2}} \frac{1}{p_1 p_2} \ge \sum_{\substack{p_1, p_2 \le \sqrt{N} \\ p_1 \ne p_2}} \frac{1}{p_1 p_2}$.

Exercise 8.2. Let $\Omega(n)$ denote the number of prime divisors of n, counted with repetitions. Thus, if the prime factorization of n is given by $n = \prod_{i=1}^m p_i^{k_i}$, then $\omega(n) = m$, but $\Omega(n) = \sum_{i=1}^m k_i$. Use the method of proof in Theorem 8.2 to prove that

$$E_N \Omega = \frac{1}{N} \sum_{n=1}^{N} \Omega(n) = \log\log N + O(1), \text{ as } N \to \infty.$$

Exercise 8.3. Let $d(n)$ denote the number of divisors of n. Thus, $d(12) = 6$ because the divisors of 12 are 1,2,3,4,6,12. Show that

$$\frac{1}{n} \sum_{j=1}^{n} d(j) = \log n + O(1).$$

This shows that the average order of the divisor function is the function $\log n$. Recall from the remark after Theorem 8.2 that the average order of $\omega(n)$, the function counting the number of distinct prime divisors, is the function $\log\log n$. (Hint: We have $d(k) = \sum_{m|k} 1$, so $\sum_{k=1}^n d(k) = \sum_{k\in[n]} \sum_{m|k} 1 = \sum_{m\in[n]} \sum_{k\in[n]:m|k} 1$.)

Chapter Notes

The theorem of G. H. Hardy and S. Ramanujan was proved in 1917. The proof we give is along the lines of the 1934 proof of P. Turán, which is much simpler than the original proof. For more on multiplicative number theory and primes, the subject of the material in Chaps. 6–8, the reader is referred to Nathanson's book [27] and to the more advanced treatment of Tenenbaum in [33]. In [27] one can find a proof of the prime number theorem by "elementary" methods. For very accessible books on analytic number theory and a proof of the prime number theorem using analytic function theory, see, for example, Apostol's book [5] or Jameson's book [25]. For a somewhat more advanced treatment, see the book of Montgomery and Vaughan [26]. One can also find a proof of the prime number theorem using analytic function theory, as well as a whole trove of sophisticated material, in [33].

Chapter 9
The Largest Clique in a Random Graph and Applications to Tampering Detection and Ramsey Theory

9.1 Graphs and Random Graphs: Basic Definitions

A *finite graph* G is a pair (V, E), where V is a finite set of *vertices* and E is a subset of $V^{(2)}$, the set of unordered pairs of elements of V. The elements of E are called *edges*. (This is what graph theorists call a *simple graph*. That is, there are no loops—edges connecting a vertex to itself—and there are no multiple edges, more than one edge connecting the same pair of vertices.) If $x, y \in V$ and the pair $\{x, y\} \in E$, then we say that an edge joins the vertices x and y; otherwise, we say that there is no edge joining x and y. If $|V| = n$, then $|V^{(2)}| = \binom{n}{2} = \frac{1}{2}n(n-1)$. The *size* of the graph is the number of vertices it contains, that is, $|V|$. We will identify the vertex set V of a graph of size n with $[n]$. The graph $G = (V, E)$ with $|V| = n$ and $E = V^{(2)}$ is called the *complete graph* of size n and is henceforth denoted by K_n. This graph has n vertices and an edge connects every one of the $\frac{1}{2}n(n-1)$ pairs of vertices. See Fig. 9.1.

For a graph $G = (V, E)$ of size n, a *clique* of size $k \in [n]$ is a complete subgraph K of G of size k; that is, $K = (V_K, E_K)$, where $V_K \subset V$, $|V_K| = k$ and $E_K = V_K^{(2)}$. See Fig. 9.2.

Consider the vertex set $V = [n]$. Now construct the edge set $E \subset [n]^{(2)}$ in the following *random* fashion. Let $p \in (0, 1)$. For each pair $\{x, y\} \in [n]^{(2)}$, toss a coin with probability p of heads and $1-p$ of tails. If heads occurs, include the pair $\{x, y\}$ in E, and if tails occurs, do not include it in E. Do this independently for every pair $\{x, y\} \in [n]^{(2)}$. Denote the resulting random edge set by $E_n(p)$. The resulting *random graph* is sometimes called an *Erdős–Rényi graph*; it will be denoted by $G_n(p) = ([n], E_n(p))$. In this chapter, the generic notation P for probability and E for expectation will be used throughout.

To get a feeling for how many edges one expects to see in the random graph, attach to each of the $N := \frac{1}{2}n(n-1)$ potential edges a random variable which is equal to 1 if the edge exists in the random set of edges $E_n(p)$ and is equal to 0 if the edge does not exist in $E_n(p)$. Denote these random variables by $\{W_m\}_{m=1}^N$. The random variables are distributed according to the Bernoulli distribution with

DOI 10.1007/978-3-319-07965-3_9, © Springer International Publishing Switzerland 2014

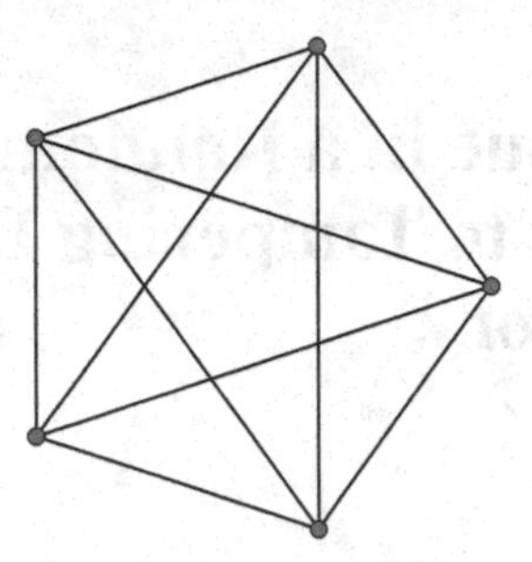

Fig. 9.1 The complete graph with 5 vertices, $G = K_5$

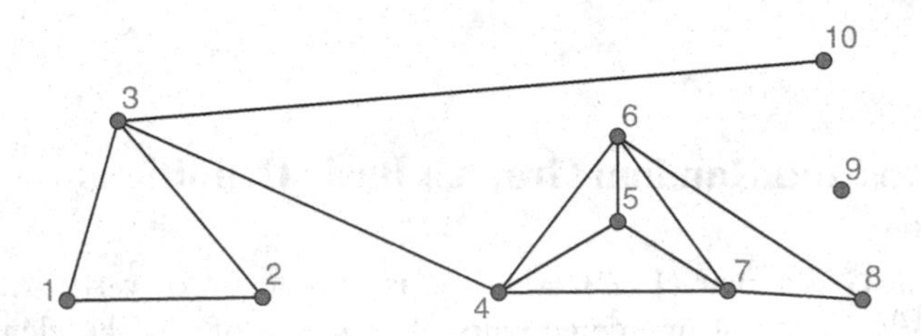

Fig. 9.2 A graph with 10 vertices and 13 edges. The largest clique is the one of size 4, formed by the vertices $\{4, 5, 6, 7\}$

parameter p; that is, $P(W_m = 1) = 1 - P(W_m = 0) = p$. Thus, the expectation and the variance of W_m are given by $EW_m = p$ and $\sigma^2(W_m) = p(1-p)$. Let $S_N = \sum_{m=1}^{N} W_m$ denote the number of edges in the random graph. By the linearity of the expectation, one has $ES_N = Np$. Because edges have been selected independently, the random variables $\{W_m\}_{m=1}^{N}$ are independent. Thus, the variance of S_N is the sum of the variances of $\{W_m\}_{m=1}^{N}$; that is, $\sigma^2(S_N) = Np(1-p)$. Therefore, Chebyshev's inequality gives

$$P(|S_N - Np| \geq N^{\frac{1+\epsilon}{2}}) \leq \frac{Np(1-p)}{N^{1+\epsilon}}.$$

Consequently, for any $\epsilon > 0$, one has $\lim_{N\to\infty} P(|S_N - Np| \geq N^{\frac{1+\epsilon}{2}}) = 0$. Thus, for any $\epsilon > 0$ and large n (depending on ϵ), with high probability the Erdős–Rényi graph $G_n(p)$ will have $\frac{1}{2}n^2 p + O(n^{1+\epsilon})$ edges.

The main question we address in this chapter is this: *how large is the largest complete subgraph, that is, the largest clique, in $G_n(p)$, as $n \to \infty$?* We study this question in Sect. 9.2. In Sect. 9.3 we apply the results of Sect. 9.2 to a problem in *tampering detection*. In Sect. 9.4, we discuss *Ramsey theory* for cliques in graphs and use random graphs to give a bound on the size of a fundamental deterministic quantity.

9.2 The Size of the Largest Clique in a Random Graph

Let $L_{n,p}$ be the random variable denoting the size of the largest clique in $G_n(p)$. Let $\log^{(2)}_{\frac{1}{p}} n := \log_{\frac{1}{p}} \log_{\frac{1}{p}} n$.

Theorem 9.1. *Let $L_{n,p}$ denote the size of the largest clique in the Erdős–Rényi graph $G_n(p)$. Then*

$$\lim_{n\to\infty} P\big(L_{n,p} \geq 2\log_{\frac{1}{p}} n - c\log^{(2)}_{\frac{1}{p}} n\big) = \begin{cases} 0, & \text{if } c < 2; \\ 1, & \text{if } c > 2. \end{cases}$$

Remark. Despite the increasing randomness and disorder in $G_n(p)$ as n grows, the theorem shows that $L_{n,p}$ behaves almost deterministically—with probability approaching 1 as $n \to \infty$, the size of the largest clique will be very close to $2\log_{\frac{1}{p}} n - 2\log^{(2)}_{\frac{1}{p}} n$. In fact, it is known that for each n, there exists a value d_n such that $\lim_{n\to\infty} P(L_n$ equals either d_n or $d_n + 1) = 1$. That is, with probability approaching 1 as $n \to \infty$, L_n is restricted to two specific values. The proof of this is similar to the proof of Theorem 9.1 but a little more delicate; see [9]. We have chosen the formulation in Theorem 9.1 in particular because it is natural for the topic discussed in Sect. 9.3.

Let $N_{n,p}(k)$ be the random variable denoting the *number* of cliques of size k in the random graph $G_n(p)$. We will always assume tacitly that the argument of $N_{n,p}$ is a positive integer. Of course it follows from Theorem 9.1 that $\lim_{n\to\infty} P(N_{n,p}(k_n) = 0) = 1$, if $k_n \geq 2\log_{\frac{1}{p}} n - c\log^{(2)}_{\frac{1}{p}} n$, for some $c < 2$. We say then that the random variable $N_{n,p}(k_n)$ *converges in probability to 0 as* $n \to \infty$. The proof of Theorem 9.1 will actually show that if $k_n \leq 2\log_{\frac{1}{p}} n - c\log^{(2)}_{\frac{1}{p}} n$, for some $c > 2$, then $\lim_{n\to\infty} P(N_{n,p}(k_n) > M) = 1$, for any $M \in \mathbb{R}$. We say then that the random variable $N_{n,p}(k_n)$ *converges in probability to* ∞ *as* $n \to \infty$. We record this as a corollary.

Corollary 9.1.

i. If $k_n \geq 2\log_{\frac{1}{p}} n - c\log^{(2)}_{\frac{1}{p}} n$, for some $c < 2$, then $N_{n,p}(k_n)$ converges to 0 in probability; that is,

$$\lim_{n\to\infty} P(N_{n,p}(k_n) = 0) = 1;$$

ii. If $k_n \leq 2\log_{\frac{1}{p}} n - c\log^{(2)}_{\frac{1}{p}} n$, for some $c > 2$, then $N_{n,p}(k_n)$ converges to ∞ in probability; that is,

$$\lim_{n\to\infty} P(N_{n,p}(k_n) > M) = 1, \text{ for all } M \in \mathbb{R}.$$

Proof of Theorem 9.1. The number of cliques of size k_n in the complete graph K_n is $\binom{n}{k_n}$; denote these cliques by $\{K_j^n : j = 1, \ldots, \binom{n}{k_n}\}$. Let $I_{K_j^n}$ be the indicator random variable defined to be equal to 1 or 0, according to whether the clique K_j^n is or is not contained in the random graph $G_n(p)$. Then we can represent the random variable $N_{n,p}(k_n)$, denoting the number of cliques of size k_n in the random graph $G_n(p)$, as

$$N_{n,p}(k_n) = \sum_{j=1}^{\binom{n}{k_n}} I_{K_j^n}. \tag{9.1}$$

Let $P(K_j^n)$ denote the probability that the clique K_j^n is contained in $G_n(p)$; that is, the probability that the edges of the clique K_j^n are all contained in the random edge set $E_n(p)$ of $G_n(p)$. Since each clique K_j^n contains $\binom{k_n}{2}$ edges, we have

$$P(K_j^n) = p^{\binom{k_n}{2}}.$$

The expected value $EI_{K_j^n}$ of $I_{K_j^n}$ is given by $EI_{K_j^n} = P(K_j^n)$. Thus, the expected value of $N_{n,p}(k_n)$ is given by

$$EN_{n,p}(k_n) = \sum_{j=1}^{\binom{n}{k_n}} EI_{K_j^n} = \binom{n}{k_n} p^{\binom{k_n}{2}}. \tag{9.2}$$

We will first prove that if $c < 2$, then

$$\lim_{n\to\infty} P(L_{n,p} \geq 2\log_{\frac{1}{p}} n - c\log^{(2)}_{\frac{1}{p}} n) = 0. \tag{9.3}$$

We have

$$EN_{n,p}(k_n) \geq P(N_{n,p}(k_n) \geq 1) = P(L_{n,p} \geq k_n),$$

where the equality follows from the fact that a clique of size l contains sub-cliques of size j for all $j \in [l-1]$. Thus, to prove (9.3) it suffices to prove that

$$\lim_{n\to\infty} EN_{n,p}(2\log_{\frac{1}{p}} n - c_n\log^{(2)}_{\frac{1}{p}} n) = 0, \tag{9.4}$$

where $0 \leq c_n \leq c < 2$, for all n. (We have written c_n instead of c in (9.4) because we need the argument of $N_{n,p}$ to be an integer.) This approach to proving (9.3) is known as the *first moment method*.

To prove (9.4), we need the following lemma.

Lemma 9.1. *If $k_n = o(n^{\frac{1}{2}})$, as $n \to \infty$, then*

$$\binom{n}{k_n} \sim \frac{n^{k_n}}{k_n!}, \text{ as } n \to \infty.$$

Proof. We have $\binom{n}{k_n} = \frac{n(n-1)\cdots(n-k_n+1)}{k_n!}$. Thus, to prove the lemma we need to show that

$$\lim_{n\to\infty} \frac{n(n-1)\cdots(n-k_n+1)}{n^{k_n}} = 1,$$

or, equivalently,

$$\lim_{n\to\infty} \sum_{j=1}^{k_n-1} \log(1-\frac{j}{n}) = 0. \tag{9.5}$$

Letting $f(x) = -\log(1-x)$, and applying Taylor's remainder theorem in the form $f(x) = f(0) + f'(x^*(x))x$, for $x > 0$, where $x^*(x) \in (0,x)$, we have

$$0 \le -\log(1-x) \le 2x,\ 0 \le x \le \frac{1}{2}.$$

Thus, for n sufficiently large so that $\frac{k_n}{n} \le \frac{1}{2}$, we have

$$0 \le -\sum_{j=1}^{k_n-1} \log(1-\frac{j}{n}) \le 2\sum_{j=1}^{k_n-1} \frac{j}{n} = \frac{(k_n-1)k_n}{n}.$$

Letting $n \to \infty$ in the above equation, and using the assumption that $k_n = o(n^{\frac{1}{2}})$, we obtain (9.5). □

We can now prove (9.4). Let $k_n = 2\log_{\frac{1}{p}} n - c_n \log^{(2)}_{\frac{1}{p}} n$, where $0 \le c_n \le c < 2$, for all n. Stirling's formula gives

$$k_n! \sim k_n^{k_n} e^{-k_n} \sqrt{2\pi k_n}, \text{ as } n \to \infty.$$

Using this with Lemma 9.1 and (9.2), we have

$$EN_{n,p}(k_n) = \binom{n}{k_n} p^{\binom{k_n}{2}} \sim \frac{n^{k_n}}{k_n!} p^{\frac{k_n(k_n-1)}{2}} \sim \frac{n^{k_n} p^{\frac{k_n(k_n-1)}{2}}}{k_n^{k_n} e^{-k_n}\sqrt{2\pi k_n}}, \text{ as } n \to \infty,$$

and thus

$$\log_{\frac{1}{p}} EN_{n,p}(k_n) = \log_{\frac{1}{p}} \binom{n}{k_n} p^{\binom{k_n}{2}} \sim$$
$$k_n \log_{\frac{1}{p}} n - \frac{1}{2}k_n^2 + \frac{1}{2}k_n - k_n \log_{\frac{1}{p}} k_n + k_n \log_{\frac{1}{p}} e - \frac{1}{2}\log_{\frac{1}{p}} 2\pi k_n, \text{ as } n \to \infty. \tag{9.6}$$

Note that

$$\log_{\frac{1}{p}} k_n = \log_{\frac{1}{p}}(2\log_{\frac{1}{p}} n - c_n \log^{(2)}_{\frac{1}{p}} n) = \log_{\frac{1}{p}}\left((\log_{\frac{1}{p}} n)(2 - \frac{c_n \log^{(2)}_{\frac{1}{p}} n}{\log_{\frac{1}{p}} n})\right) =$$
$$\log^{(2)}_{\frac{1}{p}} n + \log_{\frac{1}{p}}(2 - \frac{c_n \log^{(2)}_{\frac{1}{p}} n}{\log_{\frac{1}{p}} n}) = \log^{(2)}_{\frac{1}{p}} n + O(1), \text{ as } n \to \infty. \tag{9.7}$$

Substituting for k_n and using (9.7), we have

$$k_n \log_{\frac{1}{p}} n - \frac{1}{2}k_n^2 - k_n \log_{\frac{1}{p}} k_n = (2\log_{\frac{1}{p}} n - c_n \log^{(2)}_{\frac{1}{p}} n)\log_{\frac{1}{p}} n -$$
$$\frac{1}{2}(2\log_{\frac{1}{p}} n - c_n \log^{(2)}_{\frac{1}{p}} n)^2 - (2\log_{\frac{1}{p}} n - c_n \log^{(2)}_{\frac{1}{p}} n)(\log^{(2)}_{\frac{1}{p}} n + O(1)) =$$
$$(c_n - 2)(\log_{\frac{1}{p}} n)\log^{(2)}_{\frac{1}{p}} n + O\big(\log_{\frac{1}{p}} n\big). \tag{9.8}$$

Since $\frac{1}{2}k_n + k_n \log_{\frac{1}{p}} e - \frac{1}{2}\log_{\frac{1}{p}} 2\pi k_n = O(\log_{\frac{1}{p}} n)$, it follows from (9.6), (9.8), and the fact that $0 \le c_n \le c < 2$ that

$$\lim_{n\to\infty} \log_{\frac{1}{p}} EN_{n,p}(2\log_{\frac{1}{p}} n - c_n \log^{(2)}_{\frac{1}{p}} n) = -\infty.$$

Thus, (9.4) holds, completing the proof of (9.3).

We now prove that if $c > 2$, then

$$\lim_{n\to\infty} P(L_{n,p} \ge 2\log_{\frac{1}{p}} n - c\log^{(2)}_{\frac{1}{p}} n) = 1. \tag{9.9}$$

The analysis in the above paragraph shows that if $c_n \ge c > 2$, for all n, then

$$\lim_{n\to\infty} EN_{n,p}(2\log_{\frac{1}{p}} n - c_n \log^{(2)}_{\frac{1}{p}} n) = \infty. \tag{9.10}$$

The first moment method used above exploits the fact that (9.4) implies (9.3). Now (9.10) does not imply (9.9). To prove (9.9), we employ the second moment method. (This method was also used in Chap. 3 and Chap. 8.) The variance of $N_{n,p}(k_n)$ is given by

$$\text{Var}\big(N_{n,p}(k_n)\big) = E\Big(N_{n,p}(k_n) - EN_{n,p}(k_n)\Big)^2 = EN^2_{n,p}(k_n) - \big(EN_{n,p}(k_n)\big)^2. \tag{9.11}$$

Our goal now is to show that if $k_n = 2\log_{\frac{1}{p}} n - c_n \log^{(2)}_{\frac{1}{p}} n$ with $c_n \geq c > 2$, for all n, then

$$\text{Var}\big(N_{n,p}(k_n)\big) = o\Big(\big(EN_{n,p}(k_n)\big)^2\Big), \text{ as } n \to \infty. \tag{9.12}$$

Chebyshev's inequality gives for any $\epsilon > 0$

$$P\big(|N_{n,p}(k_n) - EN_{n,p}(k_n)| \geq \epsilon |EN_{n,p}(k_n)|\big) \leq \frac{\text{Var}\big(N_{n,p}(k_n)\big)}{\epsilon^2\big(EN_{n,p}(k_n)\big)^2}. \tag{9.13}$$

Thus, (9.12) and (9.13) yield

$$\lim_{n\to\infty} P(|\frac{N_{n,p}(k_n)}{EN_{n,p}(k_n)} - 1| < \epsilon) = 1, \text{ for all } \epsilon > 0. \tag{9.14}$$

From (9.14) and (9.10), it follows that

$$\lim_{n\to\infty} P(N_{n,p}(k_n) > M) = 1, \text{ for all } M \in \mathbb{R}. \tag{9.15}$$

In particular then, (9.9) follows from (9.15). Thus, the proof of the theorem will be complete when we prove (9.12), or, in light of (9.11), when we prove that

$$EN^2_{n,p}(k_n) = \big(EN_{n,p}(k_n)\big)^2 + o\Big(\big(EN_{n,p}(k_n)\big)^2\Big), \text{ as } n \to \infty. \tag{9.16}$$

We relabel the cliques $\{K^n_j : j = 1, \ldots, \binom{n}{k_n}\}$, of size k_n in K_n according to the vertices that are contained in each clique. Thus, we write $K^n_{i_1,i_2,\ldots,i_{k_n}}$ to denote the clique whose vertices are $i_1, i_2, \ldots, i_{k_n}$. The representation for $N_{n,p}(k_n)$ in (9.1) becomes

$$N_{n,p}(k_n) = \sum_{1\leq i_1<i_2<\cdots<i_{k_n}\leq n} I_{K^n_{i_1,i_2,\ldots,i_{k_n}}}. \tag{9.17}$$

Note that the random variable $I_{K^n_{i_1,i_2,\ldots,i_{k_n}}} I_{K^n_{l_1,l_2,\ldots,l_{k_n}}}$ is equal to 1 if the edges of the two cliques $K^n_{i_1,i_2,\ldots,i_{k_n}}$ and $K^n_{l_1,l_2,\ldots,l_{k_n}}$ are all contained in $G_n(p)$ and is equal to 0 otherwise. Thus,

$$EI_{K^n_{i_1,i_2,\ldots,i_{k_n}}} I_{K^n_{l_1,l_2,\ldots,l_{k_n}}} = P(K^n_{i_1,i_2,\ldots,i_{k_n}} \cup K^n_{l_1,l_2,\ldots,l_{k_n}}),$$

where $P(K^n_{i_1,i_2,\ldots,i_{k_n}} \cup K^n_{l_1,l_2,\ldots,l_{k_n}})$ is the probability that the edges of $K^n_{i_1,i_2,\ldots,i_{k_n}}$ and $K^n_{l_1,l_2,\ldots,l_{k_n}}$ are all contained in the random edge set $E_n(p)$ of $G_n(p)$. Consequently, we have

$$EN^2_{n,p}(k_n) = \sum_{\substack{1\le i_1<i_2<\cdots<i_{k_n}\le n\\ 1\le l_1<l_2<\cdots<l_{k_n}\le n}} EI_{K^n_{i_1,i_2,\ldots,i_{k_n}}} I_{K^n_{l_1,l_2,\ldots,l_{k_n}}} =$$

$$\sum_{\substack{1\le i_1<i_2<\cdots<i_{k_n}\le n\\ 1\le l_1<l_2<\cdots<l_{k_n}\le n}} P(K^n_{i_1,i_2,\ldots,i_{k_n}} \cup K^n_{l_1,l_2,\ldots,l_{k_n}}). \tag{9.18}$$

Now by symmetry considerations, it follows that the sum

$$\sum_{1\le l_1<l_2<\cdots<l_{k_n}\le n} P(K^n_{i_1,i_2,\ldots,i_{k_n}} \cup K^n_{l_1,l_2,\ldots,l_{k_n}})$$

over all k_n-tuples $1 \le l_1 < l_2 < \cdots < l_{k_n} \le n$ is *independent* of the particular choice of k_n-tuple $i_1, i_2, \ldots, i_{k_n}$. (The reader should verify this.) For convenience, we select the k_n-tuple $1, 2, \ldots, k_n$. Since there are $\binom{n}{k_n}$ different k_n-tuples, we have

$$EN^2_{n,p}(k_n) = \binom{n}{k_n} \sum_{1\le l_1<l_2<\cdots<l_{k_n}\le n} P(K^n_{1,2,\ldots,k_n} \cup K^n_{l_1,l_2,\ldots,l_{k_n}}). \tag{9.19}$$

Let

$$J = J(l_1, l_2, \ldots, l_{k_n}) = |[k_n] \cap \{l_1, l_2, \ldots, l_{k_n}\}|$$

denote the number of vertices shared by the cliques $K^n_{1,2,\ldots,k_n}$ and $K^n_{l_1,l_2,\ldots,l_{k_n}}$. Each of these two cliques has $\binom{k_n}{2}$ edges. Since the cliques share J vertices, the number of edges in $K^n_{1,2,\ldots,k_n} \cup K^n_{l_1,l_2,\ldots,l_{k_n}}$ is equal to $2\binom{k_n}{2} - \binom{J}{2}$, if $J \ge 2$, and is equal to $2\binom{n}{k_n}$, if $J = 0$ or $J = 1$. Thus,

$$P(K^n_{1,2,\ldots,k_n} \cup K^n_{l_1,l_2\ldots,l_{k_n}}) = \begin{cases} p^{2\binom{k_n}{2}-\binom{J}{2}}, & \text{if } J = J(l_1, l_2, \ldots, l_{k_n}) \ge 2; \\ p^{2\binom{k_n}{2}}, & \text{if } J = J(l_1, l_2, \ldots, l_{k_n}) \le 1. \end{cases} \tag{9.20}$$

Substituting (9.20) into (9.19), we have

$$EN^2_{n,p}(k_n) = \binom{n}{k_n} \sum_{\substack{1\le l_1<l_2<\cdots<l_{k_n}\le n\\ J(l_1,l_2,\ldots,l_{k_n})\ge 2}} p^{2\binom{k_n}{2}-\binom{J}{2}} + \binom{n}{k_n} \sum_{\substack{1\le l_1<l_2<\cdots<l_{k_n}\le n\\ J(l_1,l_2,\ldots,l_{k_n})\le 1}} p^{2\binom{k_n}{2}}. \tag{9.21}$$

Keep in mind that our aim is to prove (9.16). We will do this by showing that the first term on the right hand side of (9.21) is equal to $o\Big(\big(EN_{n,p}(k_n)\big)^2\Big)$ and that the second term on the right hand side of (9.21) is equal to $\big(EN_{n,p}(k_n)\big)^2 + o\Big(\big(EN_{n,p}(k_n)\big)^2\Big)$.

In order to analyze the two terms on the right hand side of (9.21), we need to count the number of k_n-tuples $l_1, l_2, \ldots, l_{k_n}$ for which $J(l_1, l_2, \ldots, l_{k_n}) = j$, for $j = 0, 1, \ldots, k_n$. Denote this number by $\#(j)$. In order that $J(l_1, l_2, \ldots, l_{k_n}) = j$, we need to choose j of the vertices of $l_1, l_2, \ldots, l_{k_n}$ from the set $[k_n]$ and the other $k_n - j$ vertices of $l_1, l_2, \ldots, l_{k_n}$ from the set $[n] - [k_n]$. Thus,

$$\#(j) = \binom{k_n}{j}\binom{n-k_n}{k_n-j}, \ j = 0, 1, \ldots, k_n. \tag{9.22}$$

We first show that the second term on the right hand side of (9.21) is equal to $\big(EN_{n,p}(k_n)\big)^2 + o\Big(\big(EN_{n,p}(k_n)\big)^2\Big)$. Using (9.22), we have

$$\binom{n}{k_n}\sum_{\substack{1\le l_1<l_2<\cdots<l_{k_n}\le n\\ J(l_1,l_2,\ldots,l_{k_n})\le 1}} p^{2\binom{k_n}{2}} = \binom{n}{k_n}\Big[\binom{k_n}{0}\binom{n-k_n}{k_n-0}+\binom{k_n}{1}\binom{n-k_n}{k_n-1}\Big]p^{2\binom{k_n}{2}} =$$

$$\binom{n}{k_n}\Big[\binom{n-k_n}{k_n} + k_n\binom{n-k_n}{k_n-1}\Big]p^{2\binom{k_n}{2}} = \big(EN_{n,p}(k_n)\big)^2\frac{\Big[\binom{n-k_n}{k_n}+k_n\binom{n-k_n}{k_n-1}\Big]}{\binom{n}{k_n}}, \tag{9.23}$$

where (9.2) was used for the final equality. By Lemma 9.1, $\binom{n}{k_n} \sim \frac{n^{k_n}}{k_n!}$, and applying Lemma 9.1 with n replaced by $n-k_n$, we have $\binom{n-k_n}{k_n} \sim \frac{(n-k_n)^{k_n}}{k_n!} = \frac{n^{k_n}}{k_n!}(1-\frac{k_n}{n})^{k_n} \sim \frac{n^{k_n}}{k_n!}$, since $k_n = o(n^{\frac{1}{2}})$. Of course then also $\binom{n-k_n}{k_n-1} \sim \frac{n^{k_n-1}}{(k_n-1)!}$. Thus,

$$\frac{\binom{n-k_n}{k_n}+k_n\binom{n-k_n}{k_n-1}}{\binom{n}{k_n}} \sim \frac{\frac{n^{k_n}}{k_n!}+k_n\frac{n^{k_n-1}}{(k_n-1)!}}{\frac{n^{k_n}}{k_n!}} = 1+\frac{k_n^2}{n}. \tag{9.24}$$

From (9.23) and (9.24), we conclude that the second term on the right hand side of (9.21) is equal to $\big(EN_{n,p}(k_n)\big)^2 + o\Big(\big(EN_{n,p}(k_n)\big)^2\Big)$.

Now we consider the first term on the right hand side of (9.21). Of course, $\binom{n-k_n}{k_n-j} \le \frac{n^{k_n-j}}{(k_n-j)!}$ and $\binom{k_n}{j} \le \frac{k_n^j}{j!}$. Also, by Lemma 9.1, $\binom{n}{k_n} \sim \frac{n^{k_n}}{k_n!}$. Using these estimates and (9.22), and recalling from (9.2) that $\big(EN_{n,p}(k_n)\big)^2 = \binom{n}{k_n}^2 p^{2\binom{k_n}{2}}$, we can estimate the first term on the right hand side of (9.21) by

$$\binom{n}{k_n}\sum_{\substack{1\le l_1<l_2<\cdots<l_{k_n}\le n\\ J(l_1,l_2,\ldots,l_{k_n})\ge 2}} p^{2\binom{k_n}{2}-\binom{J}{2}}=\binom{n}{k_n}\sum_{j=2}^{k_n}\binom{k_n}{j}\binom{n-k_n}{k_n-j}p^{2\binom{k_n}{2}-\binom{j}{2}}=$$

$$(EN_{n,p}(k_n))^2\sum_{j=2}^{k_n}\frac{\binom{k_n}{j}\binom{n-k_n}{k_n-j}}{\binom{n}{k_n}}p^{-\binom{j}{2}}\le(EN_{n,p}(k_n))^2\sum_{j=2}^{k_n}\frac{n^{k_n-j}k_n^j}{(k_n-j)!j!\binom{n}{k_n}}p^{-\binom{j}{2}}\sim$$

$$(EN_{n,p}(k_n))^2\sum_{j=2}^{k_n}\frac{n^{k_n-j}k_n^jk_n!}{(k_n-j)!j!n^{k_n}}p^{-\binom{j}{2}}\le(EN_{n,p}(k_n))^2\sum_{j=2}^{k_n}\frac{k_n^{2j}}{n^jj!}p^{-\frac{j(j-1)}{2}}. \tag{9.25}$$

By Stirling's formula, $j!\sim j^je^{-j}\sqrt{2\pi j}$, as $j\to\infty$, and thus there exists a constant $C>0$ such that

$$j!\ge Cj^je^{-j},\text{ for all } j\ge 2. \tag{9.26}$$

It is easy to check that $jp^{\frac{j-1}{2}}$ is decreasing in j for j sufficiently large. Using this and the fact that $\lim_{j\to\infty}jp^{\frac{j-1}{2}}=0$, it follows that

$$\min_{2\le j\le k_n} jp^{\frac{j-1}{2}}=k_np^{\frac{k_n-1}{2}},\text{ for sufficiently large } k_n. \tag{9.27}$$

Using (9.26) for the first inequality below and (9.27) for the second inequality below, for sufficiently large n the summation in the last term on the right hand side of (9.25) can be estimated by

$$\sum_{j=2}^{k_n}\frac{k_n^{2j}}{n^jj!}p^{-\frac{j(j-1)}{2}}\le\frac{1}{C}\sum_{j=2}^{k_n}\left(\frac{ek_n^2}{jnp^{\frac{j-1}{2}}}\right)^j\le\frac{1}{C}\sum_{j=2}^{k_n}\left(\frac{ek_n}{np^{\frac{k_n-1}{2}}}\right)^j\le$$

$$\frac{1}{C}\sum_{j=2}^{\infty}\left(\frac{\sqrt{p}ek_n}{np^{\frac{k_n}{2}}}\right)^j=\frac{1}{C}\frac{\rho_n^2}{1-\rho_n},\text{ if }\rho_n:=\frac{\sqrt{p}ek_n}{np^{\frac{k_n}{2}}}<1. \tag{9.28}$$

Using the fact that $k_n=2\log_{\frac{1}{p}}n-c_n\log^{(2)}_{\frac{1}{p}}n$ with $c_n\ge c>2$, we now show that

$$\lim_{n\to\infty}\rho_n=\sqrt{p}e\lim_{n\to\infty}\frac{k_n}{np^{\frac{k_n}{2}}}=0. \tag{9.29}$$

Using (9.7) (which of course holds for $\{c_n\}_{n=1}^\infty$ as above) for the second equality below, we have

$$\log_{\frac{1}{p}} \frac{k_n}{np^{\frac{k_n}{2}}} = \log_{\frac{1}{p}} k_n - \log_{\frac{1}{p}} n - \frac{k_n}{2} \log_{\frac{1}{p}} p = \log^{(2)}_{\frac{1}{p}} n + O(1) - \log_{\frac{1}{p}} n + \frac{k_n}{2} =$$

$$\log^{(2)}_{\frac{1}{p}} n + O(1) - \log_{\frac{1}{p}} n + \log_{\frac{1}{p}} n - \frac{c_n}{2} \log^{(2)}_{\frac{1}{p}} n = (1 - \frac{c_n}{2}) \log^{(2)}_{\frac{1}{p}} n + O(1),$$

as $n \to \infty$.

Since $c_n \geq c > 2$, it follows from this that $\lim_{n\to\infty} \log_{\frac{1}{p}} \frac{k_n}{np^{\frac{k_n}{2}}} = -\infty$ and consequently that (9.29) holds. From (9.25), (9.28), and (9.29) we conclude that the first term on the right hand of (9.21) is indeed $o\Big(\big(EN_{n,p}(k_n)\big)^2\Big)$. This completes the proof of Theorem 9.1. □

9.3 Detecting Tampering in a Random Graph

The tampering detection problem we discuss is intimately related to Theorem 9.1 and Corollary 9.1. Consider the random graph $G_n(p) = ([n], E_n(p))$. Of course, $E_n(p) \subset [n]^{(2)}$ is a random subset of $[n]^{(2)}$. Consider now the complete graph K_n whose edge set is $[n]^{(2)}$. Let k_n satisfy $1 \leq k_n \leq n$. There are $\binom{n}{k_n}$ different cliques of size k_n in K_n. We choose one of these $\binom{n}{k_n}$ cliques at random and "add" all of its edges to the random edge set $E_n(p)$ (of course some of these additional edges might already be in $E_n(p)$); that is, we take the union of $E_n(p)$ and the edges of the randomly chosen clique. We denote this new augmented edge set by $E_n^{\text{tam};k_n}(p)$ and denote the corresponding *tampered graph* by $G_n^{\text{tam};k_n}(p)$. See Fig. 9.3.

The question we ask is whether one can detect the tampering asymptotically as $n \to \infty$. Of course, we need to define what we mean by *detecting* the tampering. For this we need to define a distance between measures.

Consider a finite set Ω and consider probability measures μ and ν on Ω. We define the *total variation distance* between μ and ν by

$$D_{\text{TV}}(\mu, \nu) := \max_{A \subset \Omega} |\mu(A) - \nu(A)|. \tag{9.30}$$

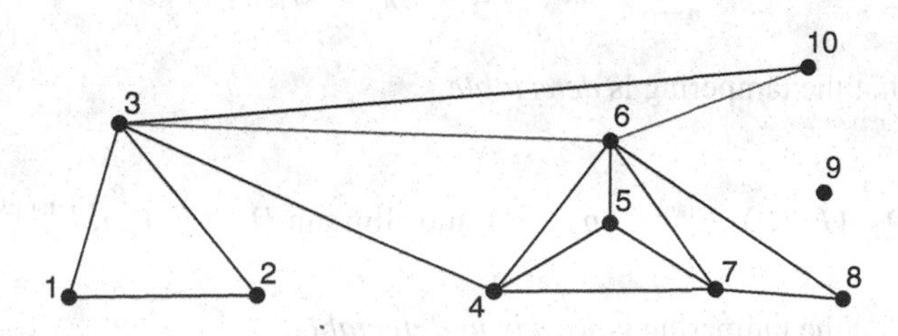

Fig. 9.3 The graph from Fig. 9.2 of size $n = 10$ has been tampered with by adding to it the clique of size $k_n = 3$ formed by the vertices $\{3,6,10\}$

In Exercise 9.1, the reader is asked to show that the distance $D_{\text{TV}}(\mu,\nu)$ can be written in two other fashions:

$$D_{\text{TV}}(\mu,\nu) = \max_{A\subset\Omega}(\mu(A) - \nu(A)) = \frac{1}{2}\sum_{x\in\Omega}|\mu(x) - \nu(x)|. \tag{9.31}$$

It is easy to see that $D_{\text{TV}}(\mu,\nu)$ takes on values in $[0,1]$, vanishes if and only if $\mu=\nu$, and equals 1 if and only if μ and ν are *mutually singular*. We recall that two probability measures μ and ν are called mutually singular if there exists a subset $A\subset\Omega$ such that $\mu(A)=\nu(\Omega-A)=1$ (and then of course $\mu(\Omega-A)=\nu(A)=0$).

Consider now a Ω-valued random variable X (defined on some probability space (S,P)). The random variable X *induces* a probability measure μ_X on Ω, namely for any subset $A\subset\Omega$, we define $\mu_X(A)=P(X\in A)$. This probability measure is called the *distribution* of X. Given two random variables X,Y taking values in Ω, we define the total variation distance between them by

$$D_{\text{TV}}(X,Y) := D_{\text{TV}}(\mu_X,\mu_Y).$$

We now apply the above concepts to the random graph. The original random graph $G_n(p)$ has as its edge set $E_n(p)$, whereas the tampered random graph $G_n^{\text{tam};k_n}(p)$ has the augmented edge set $E_n^{\text{tam};k_n}(p)$. Each of the random variables $E_n(p)$ and $E_n^{\text{tam};k_n}(p)$ takes values in the space $\mathcal{P}([n]^{(2)}) := 2^{[n]^{(2)}}$, the set of all subsets of $[n]^{(2)}$. (Given a set A, the set of all subsets of A is sometimes denoted by 2^A; it is known as the *power set* of A.) We define the tamper detection problem as follows.

Definition.

i. If

$$\lim_{n\to\infty} D_{\text{TV}}\big(E_n(p), E_n^{\text{tam};k_n}(p)\big) = 0,$$

we say that the tampering is *strongly undetectable*.

ii. If

$$\lim_{n\to\infty} D_{\text{TV}}\big(E_n(p), E_n^{\text{tam};k_n}(p)\big) = 1,$$

we say that the tampering is *detectable*.

iii. If

$$\liminf_{n\to\infty} D_{\text{TV}}(E_n(p), E_n^{\text{tam};k_n}(p)) > 0 \text{ and } \limsup_{n\to\infty} D_{\text{TV}}(E_n(p), E_n^{\text{tam};k_n}(p)) < 1,$$

we say that the tampering is *weakly undetectable*.

We will prove the following theorem.

Theorem 9.2. *Consider the Erdős–Rényi graph $G_n(p)$ with random edge set $E_n(p)$ and consider the tampered graph $G_n^{tam;k_n}(p)$ obtained by choosing at random a clique of size k_n from the complete graph K_n and adjoining its edges to $E_n(p)$ to create the augmented edge set $E_n^{tam;k_n}(p)$.*

i. If $k_n \ge 2\log_{\frac{1}{p}} n - c\log_{\frac{1}{p}}^{(2)} n$, for some $c < 2$, then the tampering is detectable; that is, $\lim_{n\to\infty} D_{TV}(E_n(p), E_n^{tam;k_n}(p)) = 1$.
ii. If $k_n \le 2\log_{\frac{1}{p}} n - c\log_{\frac{1}{p}}^{(2)} n)$, for some $c > 2$, then the tampering is strongly undetectable; that is, $\lim_{n\to\infty} D_{TV}(E_n(p), E_n^{tam;k_n}(p)) = 0$.

Remark. In light of Corollary 9.1, Theorem 9.2 seems quite intuitive. Indeed, if $k_n \ge 2\log_{\frac{1}{p}} n - c\log_{\frac{1}{p}}^{(2)} n$, with $c < 2$, then $N_{n,p}(k_n)$, the number of cliques of size k_n in the random graph $G_n(p)$, converges to 0 in probability. However, by construction, the tampered graph will always have such a clique. Thus, clearly, one can distinguish between the corresponding measures. On the other hand if $k_n \le 2\log_{\frac{1}{p}} n - c\log_{\frac{1}{p}}^{(2)} n$, with $c > 2$, then $N_{n,p}(k_n)$ converges to ∞ in probability. That is, for arbitrary M, the number of cliques of size k_n in $G_n(p)$ will be larger than M with probability approaching 1 as $n \to \infty$. Since the tampered graph $G_n^{\mathrm{tam};k_n}(p)$ is obtained from the original graph $G_n(p)$ by adjoining a randomly chosen clique of size k_n from the complete graph K_n, and since the number of cliques of size k_n in $G_n(p)$ grows unboundedly as $n \to \infty$ with probability approaching 1, it seems intuitive that the addition of a single randomly chosen clique would hardly be felt, and that asymptotically, the two graphs would be indistinguishable. Despite the above intuition, which leads to the correct answer in the present situation, there are situations in which this intuition leads one astray. See the notes at the end of the chapter.

Proof. For notational clarity at a certain point in the proof, we will denote $N_{n,p}(k_n)$, the random variable denoting the number of cliques of size k_n in the random graph $G_n(p)$, by $N_{n,p}^{(k_n)}$. For the proof of part (ii) of the theorem, we will need the weak law of large numbers for the random variable $N_{n,p}^{(k_n)}$:

$$\text{If } k_n \le 2\log_{\frac{1}{p}} n - c\log_{\frac{1}{p}}^{(2)} n, \text{ for some } c > 2, \text{ then for all } \epsilon > 0,$$
$$\lim_{n\to\infty} P(|\frac{N_{n,p}^{(k_n)}}{EN_{n,p}^{(k_n)}} - 1| < \epsilon) = 1. \tag{9.32}$$

This result was actually proved in the course of the proof of Theorem 9.1—it appears as (9.14).

Let μ_n denote the distribution of the random variable $E_n(p)$ and let $\mu_{n;\mathrm{tam}}$ denote the distribution of the random variable $E_n^{\mathrm{tam};k_n}(p)$. Let $\{K_j^n : j = 1, \ldots, \binom{n}{k_n}\}$

denote the $\binom{n}{k_n}$ cliques of size k_n in the complete graph K_n. Recall that $\mathcal{P}([n]^{(2)})$ denotes the set of subsets of $[n]^{(2)}$; thus, a point $\omega \in \mathcal{P}([n]^{(2)})$ is a subset of $[n]^{(2)}$, while a subset $A \subset \mathcal{P}([n]^{(2)})$ is a collection of subsets of $[n]^{(2)}$. Denote by $A_j^n \subset \mathcal{P}([n]^{(2)})$ the subset of $\mathcal{P}([n]^{(2)})$ consisting of all those subsets of $[n]^{(2)}$ which contain all of the $\binom{k_n}{2}$ edges of the clique K_j. Let $A^n = \cup_{j=1}^{k_n} A_j^n \subset \mathcal{P}([n]^{(2)})$ denote the set of all those subsets of $[n]^{(2)}$ which possess at least one clique of size k_n. The tampered graph is obtained by choosing at random one of the $\binom{n}{k_n}$ cliques of size k_n in K_n and adding all of its edges to the original random edge set $E_n(p)$. That is, one of the K_j^n, $j = 1, \ldots, \binom{n}{k_n}$ is chosen at random, and its edges are adjoined to $E_n(p)$ to form $E_n^{\text{tam};k_n}(p)$. Of course then, by construction, the tampered edge set $E_n^{\text{tam};k_n}(p)$ must possess a clique of size k_n; thus,

$$E_n^{\text{tam};k_n}(p) \in A^n. \tag{9.33}$$

We first prove part (i) of the theorem. Let $k_n \geq 2\log_{\frac{1}{p}} n - c\log_{\frac{1}{p}}^{(2)} n$, for some $c < 2$. By Corollary 9.1 (or Theorem 9.1), the probability of there being at least one clique of size k_n in $E_n(p)$ converges to 0 as $n \to \infty$; thus,

$$\lim_{n\to\infty} \mu_n(A^n) = 0.$$

On the other hand, by (9.33),

$$\mu_{n;\text{tam}}(A^n) = 1, \text{ for all } n.$$

Consequently,

$$D_{\text{TV}}(E_n(p), E_n^{\text{tam};k_n}(p)) = D_{\text{TV}}(\mu_n, \mu_{n;\text{tam}}) = \max_{A\subset\mathcal{P}([n]^{(2)})} |\mu_n(A) - \mu_{n;\text{tam}}(A)| \geq$$

$$|\mu_n(A^n) - \mu_{n;\text{tam}}(A^n)| = 1 - \mu_n(A^n) \to 1, \text{ as } n \to \infty,$$

proving part (i).

We now prove part (ii). The *conditional* μ_n*-probability* that a set $A \subset \mathcal{P}([n]^{(2)})$ occurs given that the set $A_j^n \subset \mathcal{P}([n]^{(2)})$ occurs is denoted by $\mu_n(A|A_j^n)$ and is given by $\mu_n(A|A_j^n) = \frac{\mu_n(A\cap A_j^n)}{\mu_n(A_j^n)}$. From the description of the construction of the tampered graph in the first paragraph of this section, along with the fact that under μ_n the existence of any particular edge is independent of the existence of any other particular edges, it follows that

$$\mu_{n;\text{tam}}(A) = \frac{1}{\binom{n}{k_n}} \sum_{j=1}^{\binom{n}{k_n}} \mu_n(A|A_j^n), \text{ for } A \subset \mathcal{P}([n]^{(2)}). \tag{9.34}$$

(The reader should verify this.)

For a point $\omega \in \mathcal{P}([n]^{(2)})$, we write $\{\omega\} \subset \mathcal{P}([n]^{(2)})$ to denote the subset of $\mathcal{P}([n]^{(2)})$ consisting of the singleton ω. Note that

$$\mu_n(\{\omega\} \cap A_j^n) = \begin{cases} \mu_n(\{\omega\}), \text{ if } \omega \in A_j^n; \\ 0, \text{ otherwise.} \end{cases}.$$

Consequently, from the definition of $N_{n,p}^{(k_n)}$ and the definition of $\{A_j^n : j = 1, \ldots, \binom{n}{k_n}\}$, we have

$$\sum_{j=1}^{\binom{n}{k_n}} \mu_n(\{\omega\} \cap A_j^n) = \mu_n(\{\omega\}) N_{n,p}^{(k_n)}(\omega), \quad \omega \in \mathcal{P}([n]^{(2)}). \tag{9.35}$$

Note that $\mu_n(A_j^n) = p^{\binom{k_n}{2}}$, for all j. Recall from (9.2) that $EN_{n,p}^{(k_n)} = \binom{n}{k_n} p^{\binom{k_n}{2}}$. Using these facts with (9.34) and (9.35), we have

$$\mu_{n;\text{tam}}(\{\omega\}) = \frac{1}{\binom{n}{k_n}} \sum_{j=1}^{\binom{n}{k_n}} \mu_n(\{\omega\}|A_j^n) = \frac{1}{\binom{n}{k_n}} \sum_{j=1}^{\binom{n}{k_n}} \frac{\mu_n(\{\omega\} \cap A_j^n)}{\mu_n(A_j^n)} =$$

$$\frac{\mu_n(\{\omega\}) N_{n,p}^{(k_n)}(\omega)}{\binom{n}{k_n} p^{\binom{k_n}{2}}} = \frac{N_{n,p}^{(k_n)}(\omega)}{EN_{n,p}^{(k_n)}} \mu_n(\{\omega\}). \tag{9.36}$$

Equation (9.36) shows that the probability measure $\mu_{n;\text{tam}}$ is the *tilted probability measure of μ_n, tilted by the random variable $N_{n,p}^{(k_n)}$*.

For $\epsilon > 0$, let

$$B_\epsilon^n = \{\omega \in \mathcal{P}([n]^{(2)}) : |\frac{N_{n,p}^{(k_n)}(\omega)}{EN_{n,p}^{(k_n)}} - 1| < \epsilon\}.$$

Since $k_n \le 2 \log_{\frac{1}{p}} n - c \log_{\frac{1}{p}}^{(2)} n$, for some $c > 2$, it follows from the law of large numbers in (9.32) that

$$\lim_{n \to \infty} \mu_n(B_\epsilon^n) = 1. \tag{9.37}$$

From (9.36), we have

$$|\mu_{n;\text{tam}}(B_\epsilon^n) - \mu_n(B_\epsilon^n)| = |\sum_{\omega \in B_\epsilon^n} \mu_n(\{\omega\})(\frac{N_{n,p}^{(k_n)}(\omega)}{EN_{n,p}^{(k_n)}} - 1)| <$$

$$\epsilon \sum_{\omega \in B_\epsilon^n} \mu_n(\{\omega\}) = \epsilon \mu_n(B_\epsilon^n) \le \epsilon, \tag{9.38}$$

where the first inequality follows from the definition of B^n_ϵ. From (9.37) and (9.38), it follows that

$$\liminf_{n\to\infty} \mu_{n:\text{tam}}(B^n_\epsilon) \ge 1 - \epsilon. \tag{9.39}$$

Now let $A \subset \mathcal{P}([n]^{(2)})$ be arbitrary. Note that (9.38) holds also with B^n_ϵ replaced by $A \cap B^n_\epsilon$; so $|\mu_{n;\text{tam}}(A \cap B^n_\epsilon) - \mu_n(A \cap B^n_\epsilon)| < \epsilon$. Let $(B^n_\epsilon)^c = \mathcal{P}([n]^{(2)}) - B^n_\epsilon$ denote the complement of B^n_ϵ. Then we have

$$\begin{aligned}
&|\mu_n(A) - \mu_{n;\text{tam}}(A)| = \\
&|\mu_n(A \cap B^n_\epsilon) + \mu_n(A \cap (B^n_\epsilon)^c) - \mu_{n;\text{tam}}(A \cap B^n_\epsilon) - \mu_{n;\text{tam}}(A \cap (B^n_\epsilon)^c)| \le \\
&|\mu_n(A \cap B^n_\epsilon) - \mu_{n;\text{tam}}(A \cap B^n_\epsilon)| + \mu_n(A \cap (B^n_\epsilon)^c) + \mu_{n;\text{tam}}(A \cap (B^n_\epsilon)^c) < \\
&\epsilon + \mu_n((B^n_\epsilon)^c) + \mu_{n;\text{tam}}((B^n_\epsilon)^c).
\end{aligned} \tag{9.40}$$

From (9.40) and the definition of the total variation distance, it follows that

$$\begin{aligned}
&D_{\text{TV}}\big(E_n(p), E_n^{\text{tam};k_n}(p)\big) = D_{\text{TV}}\big(\mu_n, \mu_{n;\text{tam}}\big) = \\
&\max_{A \subset \mathcal{P}([n]^{(2)})} |\mu_n(A) - \mu_{n;\text{tam}}(A)| < \epsilon + \mu_n((B^n_\epsilon)^c) + \mu_{n;\text{tam}}((B^n_\epsilon)^c).
\end{aligned} \tag{9.41}$$

From (9.37), (9.39), (9.41), and the fact that $\epsilon > 0$ is arbitrary, we conclude that

$$\lim_{n\to\infty} D_{\text{TV}}\big(E_n(p), E_n^{\text{tam};k_n}(p)\big) = 0. \tag{9.42}$$

□

Remark. The final two paragraphs of the proof can be replaced by a shorter argument using L^2-convergence and the Cauchy–Schwarz inequality. See Exercise 9.2.

9.4 Ramsey Theory

Consider the complete graph K_n. For each edge in K_n, choose either blue or red, and color the edge with that color. We call this a *2-coloring* of K_n. For $2 \le k \le n$, one can ask whether there exists a *monochromatic* clique of size k, that is, a clique with all of its edges blue or with all of its edges red. For $k = 2$, obviously there exists such a monochromatic clique, for all $n \ge 2$. The fundamental theorem of *Ramsey theory* states the following:
For each integer $k \ge 3$, there exists an integer $R(k) > k$ such that if $n \ge R(k)$, then every 2-coloring of K_n will necessarily have a monochromatic clique of size k, while if $k \le n < R(k)$, then it is possible to find a 2-coloring of K_n with no monochromatic clique of size k.

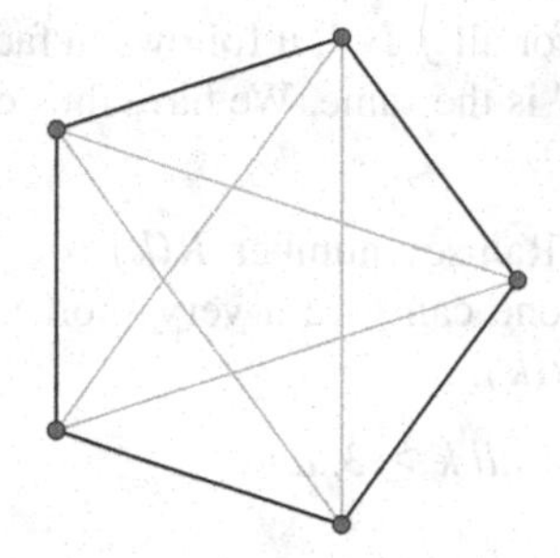

Fig. 9.4 The above example shows that $R(3) > 5$

Note that this result is purely deterministic—it says that *no matter how* we arrange the coloring of K_n, there *must* be a monochromatic clique of size k, if $n \geq R(k)$. The exact computation of the *Ramsey numbers* $R(k)$ is notoriously hard. One has $R(3) = 6$ and $R(4) = 18$, but the exact value of $R(5)$ is unknown! See Fig. 9.4.

Remark. It is known that $43 \leq R(5) \leq 49$. The complete graph K_{43} has $\frac{1}{2} \cdot 43 \cdot 42 = 903$ edges. There are 2^{903} different two-colorings of K_{43} and $\binom{43}{5} = 962,598$ different cliques of size 5.

We will prove the above fundamental result by providing upper and lower bounds on $R(k)$. A nice, elementary combinatorial argument yields the following result.

Theorem 9.3.

$$R(k) \leq 4^{k-1},\ k \geq 3. \tag{9.43}$$

Remark. The above estimate is not far from the best known asymptotic upper bound for $R(k)$. In particular, it is not known if $R(k) \leq c^k$, for large k and some $c < 4$. For the best known upper bound, see [12].

Proof. Let $k \geq 3$. Consider an arbitrary coloring of the complete graph $K_{4^{k-1}}$ of size $4^{k-1} = 2^{2k-2}$. Define $x_1 = 1$ and $S_0 = K_{4^{k-1}}$. Since x_1 shares an edge with $2^{2k-2} - 1$ vertices, there must be a set of vertices S_1 of size at least 2^{2k-3} such that every edge from x_1 to a vertex in S_1 is the same color. This is the so-called *pigeonhole principle*. Let x_2 denote the vertex in S_1 with the lowest number. By the same reasoning, since x_2 shares an edge with all the other vertices in S_1, of which there are at least $2^{2k-3} - 1$, there must be a set $S_2 \subset S_1$ of size at least 2^{2k-4} such that every edge from x_2 to a vertex in S_2 has the same color. Continuing like this, we obtain a sequence $x_1, \ldots, x_{2k-2}$ of vertices and a decreasing, nested sequence of sets of vertices $\{S_j\}_{j=0}^{2k-3}$ such that $x_j \in S_{j-1}$, $j \in [2k-2]$. By the construction, it follows that for each i, the color of the edge joining x_i to x_j is the same for all $j > i$. Now look at the $2k - 3$ edges $\{\{x_i, x_{i+1}\}\}_{i=1}^{2k-3}$. Obviously, we can choose at least $k - 1$ of these edges to be all the same color. Find such a set of edges and denote the set of vertices in these edges by S. Note that $|S| \geq k$. Because the color joining x_i to x_j is the same for all $j > i$, it follows in fact that the color of the edge joining any two vertices in S is the same. We have thus exhibited a monochromatic clique of size at least k. □

Despite the fact that the Ramsey number $R(k)$ is a quantity associated with a purely deterministic result, one can give a very short and ingenious probabilistic proof of a lower bound for $R(k)$.

Theorem 9.4. *$R(k) > k$, for all $k \geq 3$, and*

$$R(k) \geq \frac{1}{e}(1+o(1))k2^{\frac{k}{2}}, \text{ as } k \to \infty. \tag{9.44}$$

Remark. The best known lower bound is just $\sqrt{2}$ times the above estimate; see [2]. Thus, a real chasm lies between the best known upper bound and the best known lower bound!

Proof. Consider a *random* two-coloring of the graph K_n, where each edge is colored red or blue with equal probability, and independently of what occurs at other edges. Let W be a clique in K_n of size k, with $3 \leq k \leq n$. Let I_W be the indicator random variable, which is equal to 1 if W is monochromatic, and equal to 0 otherwise. Since there are $\binom{k}{2}$ edges in W, the probability that W is all blue (or all red) is $(\frac{1}{2})^{\binom{k}{2}}$; consequently, the probability that W is monochromatic is $2^{1-\binom{k}{2}}$. Of course, the expected value EI_W of I_W is also equal to $2^{1-\binom{k}{2}}$.

For $3 \leq k \leq n$, let $X_k = \sum_{|W|=k} I_W$. The random variable X_k counts the number of monochromatic cliques of size k in K_n. We have

$$EX_k = \sum_{|W|=k} EI_W = \binom{n}{k} 2^{1-\binom{k}{2}}.$$

Since the *average* number of monochromatic cliques of size k in this random two-coloring is equal to $\binom{n}{k}2^{1-\binom{k}{2}}$, there certainly must exist some particular two-coloring with exactly M monochromatic cliques of size k, for some $M \leq \binom{n}{k}2^{1-\binom{k}{2}}$. Consider such a two-coloring. From each of the M monochromatic cliques of size k, remove one of the vertices. Let M' denote the number of vertices removed. We have $M' \leq M$. (It is possible that $M' < M$ because we might have removed the same vertex from more than one of the cliques.) What remains is a two-coloring of the complete graph on $n - M'$ vertices, and by construction, this two-coloring has no monochromatic cliques of size k. We conclude that

$$R(k) > n - \binom{n}{k} 2^{1-\binom{k}{2}}, \text{ for any } n \geq k.$$

In particular, choosing $n = k + 1$, one obtains $R(k) > k + 1 - 2(k + 1)2^{-\binom{k}{2}}$, and it is easy to check that the right hand side is greater than or equal to k, for all $k \geq 3$. In Exercise 9.3 the reader is asked to show that

$$\max_{k \leq n < \infty} \left(n - \binom{n}{k} 2^{1-\binom{k}{2}}\right) = \frac{1}{e}(1 + o(1))k2^{\frac{k}{2}}, \text{ as } k \to \infty. \tag{9.45}$$

□

Remark. The strategy used to prove Theorem 9.4 is known as the *probabilistic method.* It was pioneered by P. Erdős. He used the method in a slightly different way from above and obtained a lower bound on $R(k)$ with an extra factor of $\sqrt{2}$ in the denominator on the right hand side of (9.44).

Exercise 9.1. Show that the total variation distance $D_{\text{TV}}(\mu, \nu)$ defined in (9.30) satisfies (9.31).

Exercise 9.2. This exercise presents an alternative approach in place of the final two paragraphs of the proof of part (ii) of Theorem 9.2. Recall that the *Cauchy–Schwarz* inequality states that $|\sum_{i=1}^m a_i b_i| \leq \sqrt{(\sum_{i=1}^m a_i^2)(\sum_{i=1}^m b_i^2)}$, where $\{a_i\}_{i=1}^m$, $\{b_i\}_{i=1}^m$ are real numbers and m is a positive integer.

a. Use (9.36) and the Cauchy–Schwarz inequality to show that for any $A \subset \mathcal{P}([n]^{(2)})$, one has

$$|\mu_{n;\text{tam}}(A) - \mu_n(A)| \leq \sqrt{\mu(A)} \sqrt{\sum_{\omega \in A} \Big(\frac{N_{n,p}^{(k_n)}(\omega)}{EN_{n,p}^{(k_n)}} - 1\Big)^2 \mu_n(\omega)} \leq$$

$$\sqrt{\sum_{\omega \in \mathcal{P}([n]^{(2)})} \Big(\frac{N_{n,p}^{(k_n)}(\omega)}{EN_{n,p}^{(k_n)}} - 1\Big)^2 \mu_n(\omega)}. \tag{9.46}$$

b. The expression on the right hand side of (9.46) is called the L^2-*norm with respect to the measure* μ_n of the function $\frac{N_{n,p}^{(k_n)}(\omega)}{EN_{n,p}^{(k_n)}} - 1$, which is defined on the domain $\mathcal{P}([n]^{(2)})$. We denote this norm by $||\frac{N_{n,p}^{(k_n)}}{EN_{n,p}^{(k_n)}} - 1||_{2;\mu_n}$. Use (9.16) (where the notation $N_{n,p}(k_n)$ instead of $N_{n,p}^{(k_n)}$ is used), which holds for k_n as in part (ii) of Theorem 9.2, to prove that

$$\lim_{n \to \infty} ||\frac{N_{n,p}^{(k_n)}}{EN_{n,p}^{(k_n)}} - 1||_{2;\mu_n} = 0. \tag{9.47}$$

c. Conclude from (9.46) and (9.47) that (9.42) holds.

Exercise 9.3. Show that (9.45) holds. (Hint: Let $f_{1,k}(x) = x - \frac{2^{1-\binom{k}{2}}x^k}{k!}$ and $f_{2,k}(x) = x - \frac{2^{1-\binom{k}{2}}(x-k)^k}{k!}$. Show that $\max_{k \le x < \infty} f_{1,k}(x) \sim \max_{k \le x < \infty} f_{2,k}(x)$ as $k \to \infty$. Since $\frac{(n-k)^k}{k!} \le \binom{n}{k} \le \frac{n^k}{k!}$, it then follows that $\max_{k \le n < \infty} \left(n - \binom{n}{k}2^{1-\binom{k}{2}}\right) \sim \max_{k \le x < \infty} f_{1,k}(x)$, as $k \to \infty$. To obtain the asymptotic behavior of $\max_{k \le x < \infty} f_{1,k}(x)$, you will need Stirling's formula.)

Exercise 9.4. Figure 9.4 shows that the Ramsey number $R(3)$ satisfies $R(3) \ge 5$. Prove that $R(3) = 6$.

Chapter Notes

For a wide scope of results concerning graphs, deterministic and random, see Bollobás' books [9] and [10].

For a paper that considers tampering detection, see [29]. In particular, one finds there two examples that show that the intuition for Theorem 9.2, discussed in the remark following the theorem, can fail. It should be noted that the word "detection" must be understood here in a very theoretical way, as there are no known algorithms for detecting this clique in a reasonable amount of time, namely an amount of time which grows no more than polynomially in the number of vertices n. The construction of such algorithms is known in the theoretical computer science literature as the "planted clique" problem. See, for example, the paper of Alon et al. [3], where for $p = \frac{1}{2}$ it is shown that a planted clique of order $n^{\frac{1}{2}}$ can be detected in polynomial time. (This order for the clique is of course far, far larger than the order $\log n$ for the cliques discussed in this chapter.)

The proof of the existence of the Ramsey number $R(k)$ goes back to F. Ramsey in 1930. The nice little book by Alon and Spencer [2] is devoted entirely to the probabilistic method in combinatorics. The book by Graham et al. [22] is devoted entirely to Ramsey theory.

Chapter 10
The Phase Transition Concerning the Giant Component in a Sparse Random Graph: A Theorem of Erdős and Rényi

10.1 Introduction and Statement of Results

Let $G_n(p_n) = ([n], E_n(p_n))$ denote the Erdős–Rényi graph of size n which was introduced in Chap. 9. As in Chap. 9, the generic notation P for probability and E for expectation will be used in this chapter. Note that whereas in Chap. 9 the edge probability p was fixed independent of the graph size, in this chapter the edge probability p_n will vary with n. A subset $A \subset [n]$ of the vertex set $[n]$ is called *connected* if for every $x, y \in A$, there exists a path between x and y along edges in $E_n(p_n)$. The vertex set $[n]$ is of course equal to the disjoint union of its connected components. Let C_n^{lg} be the random variable denoting the size of the largest connected component in the random graph $G_n(p_n)$. It turns out that the size of the largest connected component undergoes a striking *phase transition* as the edge probability passes from $\frac{c}{n}$ with $c < 1$ to $\frac{c}{n}$ with $c > 1$. In this chapter we will prove the following two theorems.

Theorem 10.1. *Let $p_n = \frac{c}{n}$, with $c < 1$. Then there exists a $\gamma = \gamma(c)$ such that the size C_n^{lg} of the largest connected component of $G_n(p_n)$ satisfies*

$$\lim_{n\to\infty} P(C_n^{lg} \le \gamma \log n) = 1.$$

Theorem 10.2. *Let $p_n = \frac{c}{n}$, with $c > 1$. Then there exists a unique solution $\beta = \beta(c) \in (0,1)$ to the equation $1 - e^{-cx} - x = 0$. For any $\epsilon > 0$, the size C_n^{lg} of the largest connected component of $G_n(p_n)$ satisfies*

$$\lim_{n\to\infty} P\big((1-\epsilon)\beta n \le C_n^{lg} \le (1+\epsilon)\beta n\big) = 1. \tag{10.1}$$

DOI 10.1007/978-3-319-07965-3_10, © Springer International Publishing Switzerland 2014

Furthermore, every other connected component of $G_n(p_n)$ *is of size* $O(\log n)$ *as* $n \to \infty$*; that is, letting* $C_n^{\text{2nd-lg}}$ *denote the size of the second largest component, then for some* $\gamma = \gamma(c)$,

$$\lim_{n\to\infty} P(C_n^{\text{2nd-lg}} \le \gamma \log n) = 1. \tag{10.2}$$

Remark 1. In light of (10.1) and (10.2), when $p_n = \frac{c}{n}$ with $c > 1$, the largest component is referred to as *the giant component.*

Remark 2. It follows from the above theorems that when $p_n = \frac{c}{n}$, for some $c > 0$, the probability that the graph is connected approaches 0 as $n \to \infty$. This can be proved directly far more easily than the above theorems can be proved. Indeed, in Exercise 10.3, the reader is guided through a proof of the following fact concerning *disconnected* vertices; that is vertices that are not connected to any other vertices: If $p_n = \frac{\log n + c_n}{n}$, then as $n \to \infty$, the probability of there being at least one disconnected vertex approaches 0 if $\lim_{n\to\infty} c_n = \infty$, while for any M, the probability of there being at least M disconnected vertices approaches 1 if $\lim_{n\to\infty} c_n = -\infty$. Actually, there is a totally trivial way to see that if $p_n = \frac{c}{n}$, then the probability that the graph is connected does not approach 1. Indeed, simply note that the probability that any particular vertex is disconnected is $(1-\frac{c}{n})^{n-1}$; thus as $n \to \infty$, the probability that any particular vertex is disconnected converges to e^{-c}. The above theorems and this discussion naturally elicit the question, how large must p_n be in order that the graph be connected? The answer to this was given also by Erdős and Rényi, who proved that the above threshold probability concerning whether or not the graph possesses disconnected vertices is also the threshold for connectivity: If $p_n = \frac{\log n + c_n}{n}$, then as $n \to \infty$, the probability of the graph being connected approaches 1 if $\lim_{n\to\infty} c_n = \infty$ and approaches 0 if $\lim_{n\to\infty} c_n = -\infty$. See [9].

Remark 3. If a connected component of a graph contains m vertices, then it must contain at least $m-1$ edges. Thus, it follows from (10.1) that if $c > 1$, then for any $\epsilon > 0$ and for large n, with high probability, the random graph $G_n(\frac{c}{n})$ will contain at least $(1-\epsilon)\beta(c)n$ edges. In Sect. 9.1 of Chap. 9, it was shown that for any $\epsilon > 0$, with high probability, the graph $G_n(p)$ has $\frac{1}{2}n^2p + O(n^{1+\epsilon})$ edges. The same type of analysis shows that for any $\epsilon > 0$ and large n, with high probability the graph $G_n(\frac{c}{n})$ has $\frac{1}{2}cn + O(n^{\frac{1}{2}+\epsilon})$ edges. Thus, one must have $\beta(c) \le \frac{c}{2}$, for $1 < c < 2$. See Exercise 10.1.

In Sect. 10.2 we construct the setup that will be used for the proofs of Theorems 10.1 and 10.2. In particular, we construct and analyze probabilistically an algorithm that calculates for each vertex of the graph the size of the connected component to which it belongs. In Sect. 10.3 we present a couple of basic *large deviations* estimates that will be needed for the proofs of the theorems. The results of Sects. 10.2 and 10.3 will allow for a quick proof of Theorem 10.1 in Sect. 10.4. In Sect. 10.5, we give a concise presentation of the *Galton–Watson branching process* and prove the most basic theorem of this subject, concerning the probability of extinction. This will be used for one part of the proof of Theorem 10.2, which is presented in Sect. 10.6. The proof of Theorem 10.2 requires considerably more technical work over and above that which is required for the proof of Theorem 10.1.

10.2 Construction of the Setup for the Proofs of Theorems 10.1 and 10.2

Let $x \in [n]$ be a vertex of the random graph. All the random quantities that we define below depend on x and n, but we suppress this dependence in the notation. We construct an algorithm that produces the connected component to which x belongs. We begin by calling x "alive" and calling all of the other vertices in $[n]$ "neutral." We define $Y_0 = 1$, to indicate that at the beginning there is one vertex that is alive. Each of the neutral vertices y is now observed. If there is an edge connecting x to y, that is, if $\{x, y\} \in E_n(p_n)$, then y is declared alive; if not, then y remains neutral. After every such y has been checked, we declare x to be "dead." We define Y_1 to be the new number of vertices that are alive. We also say that at time $t = 1$ there is one dead vertex. This ends the first step of the algorithm. We continue like this. If at the end of step t there are $Y_t > 0$ vertices that are alive (and t dead vertices), we begin step $t + 1$ by selecting one of the alive vertices (it doesn't matter which one) and call it z. Each of the currently neutral vertices y is now observed. If there is an edge connecting z and y, then y is declared alive; if not, then y remains neutral. After every such y has been checked, we declare z to be "dead." We define Y_{t+1} to be the new number of vertices that are alive, and we say that at time $t + 1$ there are $t + 1$ dead vertices. The process stops at the end of the step T for which $Y_T = 0$. It follows that at the end of step T, there are T dead vertices. A little thought shows that these dead vertices form the connected component to which x belongs. Thus T is the size of the connected component to which x belongs. (The reader should verify this.) See Fig. 10.1. Of course, T is a random variable since it depends on the random edge configuration $E_n(p_n)$.

For $1 \le t \le T$, define Z_t to be the number of neutral vertices that are declared alive at step t. Then from the description of the algorithm, we have for $1 \le t \le T$,

$$Y_t = Y_{t-1} + Z_t - 1. \tag{10.3}$$

Assuming that $t \le T$, at the end of step $t - 1$, there are $t - 1$ dead vertices and $Y_{t-1} > 0$ alive vertices. Thus there are $n - t - Y_{t-1} + 1$ neutral vertices. *A key feature of the above algorithm is that no pair of vertices is ever checked twice.* Consequently, for every pair of vertices that is checked, the probability of there being an edge between them is equal to p_n, independently of what occurred when checking other pairs of vertices. Thus, since Z_t counts how many of the $n - t - Y_{t-1} + 1$ neutral vertices have a common edge with the alive vertex z that has been selected for implementing step t, and since the probability of there being an edge

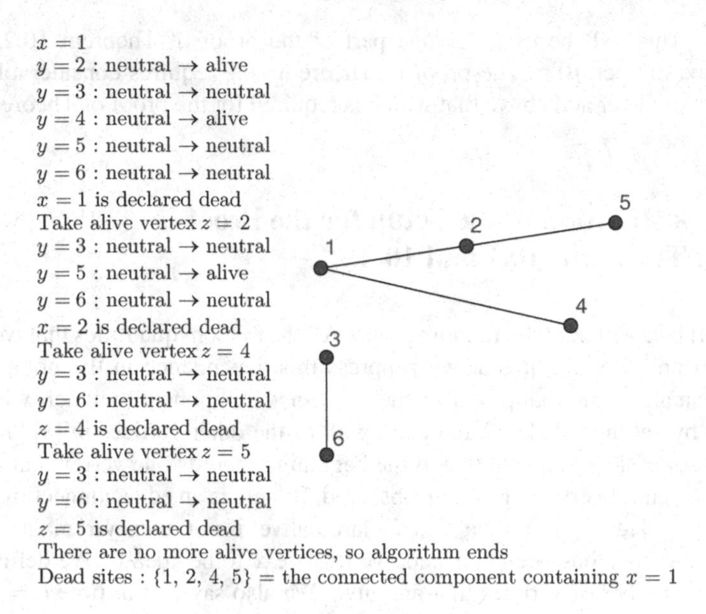

Fig. 10.1 The algorithm

from z to any given neutral vertex is p_n, it follows that Z_t is distributed according to the binomial distribution with parameters $n - t - Y_{t-1} + 1$ and p_n: for $1 \le t \le T$,

$$Z_t \sim \text{Bin}(n - t - Y_{t-1} + 1, p_n). \tag{10.4}$$

Of course, Y_{t-1}, which appears in the size parameter of the binomial distribution above, is itself a random variable. The meaning of (10.4) is that conditioned on knowing that $Y_{t-1} = y$, then $Z_t \sim \text{Bin}(n-t-y+1, p_n)$. Since no pair of vertices is ever checked twice, and since from (10.3), Y_{t-1} only depends on $\{Z_s\}_{s=1}^{t-1}$, it follows that given the value of Y_{t-1}, and given that $T \ge t$, the random variable Z_t and the random variables $\{Z_s\}_{s=1}^{t-1}$ are *conditionally independent*; that is, for all $m \ge 1$ and all $t \ge 2$,

$$P(Z_t \in \cdot, \{Z_s\}_{s=1}^{t-1} \in \cdot | Y_{t-1} = m, T \ge t) =$$
$$P(Z_t \in \cdot | Y_{t-1} = m, T \ge t) P(\{Z_s\}_{s=1}^{t-1} \in \cdot | Y_{t-1} = m, T \ge t). \tag{10.5}$$

As noted, (10.3) and (10.4) hold only up to time T; however it will be convenient to define Y_t and Z_t recursively from (10.3) and (10.4) for all integers $0 \le t \le n$. (Thus, e.g., if $T = t_0$, then we have $Y_{t_0} = 0$ (as well as $Z_{t_0} = 0$), and thus $Z_{t_0+1} \sim \text{Bin}(n - t_0, p_n)$ and $Y_{t_0+1} = Z_{t_0+1} - 1$.) In particular, for $t > T$, Y_t can take on negative values. For $1 \le t \le T$, note that the number N_t of neutral vertices at the end of step t is given by $N_t = n - t - Y_t$. We use this equation to define N_0, namely, $N_0 = n-1$, indicating that there are $n-1$ neutral vertices before the first step begins. We now use this equation to extend N_t also to all $0 \le t \le n$. We have the following key lemma.

Lemma 10.1.

$$Y_t - 1 + t \sim Bin(n - 1, 1 - (1 - p_n)^t),\ t \geq 0.$$

Proof. Since $N_t = n - t - Y_t = (n - 1) - (Y_t - 1 + t)$, the statement of the lemma is equivalent to

$$N_t \sim \text{Bin}(n - 1, (1 - p_n)^t). \tag{10.6}$$

We prove (10.6) by induction. Clearly (10.6) holds for $t = 0$. Now assume that for some $t \geq 1$,

$$N_{t-1} \sim \text{Bin}(n - 1, (1 - p_n)^{t-1}). \tag{10.7}$$

Using (10.3), we have

$$N_t = n - t - Y_t = n - t - Y_{t-1} + 1 - Z_t = N_{t-1} - Z_t. \tag{10.8}$$

However, from (10.4) and the definition of N_{t-1}, we have $Z_t \sim \text{Bin}(N_{t-1}, p_n)$. Thus, $N_{t-1} - Z_t \sim \text{Bin}(N_{t-1}, 1 - p_n)$, and it follows from (10.8) that

$$N_t \sim \text{Bin}(N_{t-1}, 1 - p_n). \tag{10.9}$$

By the inductive hypothesis (10.7), N_{t-1} is the number of heads in $n-1$ independent coin flips, where on each flip the probability of heads is $(1 - p_n)^{t-1}$. Then (10.9) states that N_t is the number of "successes" in $n - 1$ independent trials, where each trial consists of first tossing a coin with probability $(1 - p_n)^{t-1}$ of heads and then tossing a second coin with probability $1 - p_n$ of heads, and a "success" is defined as obtaining heads on both flips. This description of N_t is the description of a random variable distributed according to $\text{Bin}(n - 1, (1 - p_n)^t)$. For an alternative derivation that (10.9) and (10.7) imply (10.6), using generating functions, see Exercise 10.4. □

10.3 Some Basic Large Deviations Estimates

We present two propositions which are known as *large deviations* estimates. The first proposition will be used in the proof of Theorem 10.1 and the second one will be used in the proof of Theorem 10.2.

Proposition 10.1. *Let $c \in (0, 1)$. For $n \in \mathbb{Z}^+$ and $t > 0$ with $\frac{tc}{n} \leq 1$, let $S_{n,t} \sim Bin(n, \frac{tc}{n})$. Then there exists a $\kappa = \kappa(c) > 0$, independent of n and t, such that*

$$P(S_{n,t} \geq t) \leq e^{-\kappa t}.$$

Remark. Note that $ES_{n,t} = n(\frac{tc}{n}) = tc < t$, since $c \in (0,1)$.

Proof. For any $\lambda > 0$, we have

$$P(S_{n,t} \geq t) \leq \exp(-\lambda t) E \exp(\lambda S_{n,t}), \tag{10.10}$$

since $\exp(\lambda(S_{n,t} - t)) \geq 1$ on the event $\{S_{n,t} \geq t\}$.

Since $S_{n,t}$ is the number of successes in n independent Bernoulli trials, each of which has probability $\frac{tc}{n}$ of success, it follows that $S_{n,t}$ can be represented as $S_{n,t} = \sum_{j=1}^{n} B_j$, where the $\{B_j\}_{j=1}^{n}$ are independent and identically distributed Bernoulli random variables with parameter $\frac{tc}{n}$; that is, $P(B_j = 1) = 1 - P(B_j = 0) = \frac{tc}{n}$. Using the fact that these random variables are independent and identically distributed, we have

$$E \exp(\lambda S_{n,t}) = E \exp(\lambda \sum_{j=1}^{n} B_j) = \prod_{j=1}^{n} E \exp(\lambda B_j) = (E \exp(\lambda B_1))^n =$$

$$(1 - \frac{tc}{n} + \frac{tc}{n} e^{\lambda})^n. \tag{10.11}$$

Since $1 + y \leq e^y$, for all y, we have $(1 + \frac{x}{n})^n \leq e^x$, for all $x \geq 0$ and all $n \geq 1$. Thus, $(1 - \frac{tc}{n} + \frac{tc}{n} e^{\lambda})^n \leq e^{tc(e^{\lambda}-1)}$, and consequently, (10.10) and (10.11) give for any $\lambda > 0$

$$P(S_{n,t} \geq t) \leq e^{-\lambda t} e^{tc(e^{\lambda}-1)} = \exp\big(-(\lambda - ce^{\lambda} + c)t\big). \tag{10.12}$$

The function $f(\lambda) := \lambda - ce^{\lambda} + c$ satisfies $f(0) = 0$, and $f'(0) > 0$, since $c \in (0,1)$. Thus, there exist $\kappa = \kappa(c) > 0$ and $\lambda_0 > 0$ such that $f(\lambda_0) = \kappa$. We then conclude from (10.12) that $P(S_{n,t} \geq t) \leq e^{-\kappa t}$. □

Proposition 10.2. *For each $n \in \mathbb{Z}^+$, let $S_n \sim Bin(n, \rho)$, where $\rho \in (0,1)$. Let*

$$\kappa(\rho_0, \rho) = \rho_0 \log \frac{\rho_0}{\rho} + (1 - \rho_0) \log \frac{1 - \rho_0}{1 - \rho}, \; 0 < \rho, \rho_0 < 1.$$

Then $\kappa(\rho_0, \rho) > 0$, if $\rho \neq \rho_0$, and

(i) if $\rho < \rho_0 < 1$, then

$$P(S_n \geq \rho_0 n) \leq e^{-\kappa(\rho_0,\rho)n}, \text{ for all } n;$$

(ii) if $0 < \rho_0 < \rho$, then

$$P(S_n \leq \rho_0 n) \leq e^{-\kappa(\rho_0,\rho)n}, \text{ for all } n.$$

Remark. The function $\kappa(\rho_0, \rho)$ is a *relative entropy*. For more about this, see the notes at the end of the chapter.

Proof. The following three facts show that (ii) follows from (i): $\hat{S}_n := n - S_n$ is distributed according to the distribution $\text{Bin}(n, 1-\rho)$, $P(S_n \le \rho_0 n) = P(\hat{S}_n \ge (1-\rho_0)n)$ and $\kappa(1-\rho_0, 1-\rho) = \kappa(\rho_0, \rho)$. So it suffices to show that (i) holds and that $\kappa(\rho_0, \rho) > 0$, if $\rho \ne \rho_0$.

Let $\rho_0 > \rho$. For any $\lambda > 0$, we have

$$P(S_n \ge \rho_0 n) \le \exp(-\lambda\rho_0 n) E \exp(\lambda S_n), \tag{10.13}$$

since $\exp(\lambda(S_n - \rho_0 n)) \ge 1$ on the event $\{S_n \ge \rho_0 n\}$. We can represent the random variable S_n as $S_n = \sum_{j=1}^n B_j$, where the $\{B_j\}_{j=1}^n$ are independent and identically distributed Bernoulli random variables with parameter ρ; that is, $P(B_j = 1) = 1 - P(B_j = 0) = \rho$. Using the fact that these random variables are independent and identically distributed, we have

$$E \exp(\lambda S_n) = E \exp(\lambda \sum_{j=1}^n B_j) = \prod_{j=1}^n E \exp(\lambda B_j) = (E \exp(\lambda B_1))^n =$$

$$(\rho e^\lambda + 1 - \rho)^n. \tag{10.14}$$

Thus, from (10.13), we obtain the inequality

$$P(S_n \ge \rho_0 n) \le \left[e^{-\lambda\rho_0}(\rho e^\lambda + 1 - \rho)\right]^n, \text{ for all } n \ge 1 \text{ and all } \lambda > 0. \tag{10.15}$$

The function $f(\lambda) := e^{-\lambda\rho_0}(\rho e^\lambda + 1 - \rho)$, $\lambda \ge 0$, satisfies $f(0) = 1$, $\lim_{\lambda\to\infty} f(\lambda) = \infty$, and $f'(0) = -\rho_0 + \rho < 0$. Consequently, f possesses a global minimum at some $\lambda_0 > 0$, and $f(\lambda_0) \in (0, 1)$ [indeed, $f(\lambda_0) \le 0$ would contradict (10.15)]. In Exercise 10.5, the reader is asked to show that $f(\lambda_0) = \left(\frac{1-\rho}{1-\rho_0}\right)^{1-\rho_0}\left(\frac{\rho}{\rho_0}\right)^{\rho_0}$. Note now that $\kappa(\rho_0, \rho)$, defined in the statement of the proposition, is equal to $-\log f(\lambda_0)$. Thus, $\kappa(\rho_0, \rho) > 0$, for $\rho_0 > \rho$, and $e^{-\kappa(\rho_0,\rho)} = f(\lambda_0)$. Substituting $\lambda = \lambda_0$ in (10.15) gives

$$P(S_n \ge \rho_0 n) \le e^{-\kappa(\rho_0,\rho)n}.$$

Finally, since $\kappa(\rho_0, \rho) = \kappa(1-\rho_0, 1-\rho)$, it follows that $\kappa(\rho_0, \rho) > 0$, if $\rho \ne \rho_0$. □

10.4 Proof of Theorem 10.1

In this section, and also in Sect. 10.6, we will use tacitly the following facts, which are left to the reader in Exercise 10.6:

1. If $X_i \sim \text{Bin}(n_i, p)$, $i = 1, 2$, and $n_1 > n_2$, then $P(X_1 \geq k) \geq P(X_2 \geq k)$, for all integers $k \geq 0$.
2. If $X_i \sim \text{Bin}(n, p_i)$, $i = 1, 2$, and $p_1 > p_2$, then $P(X_1 \geq k) \geq P(X_2 \geq k)$, for all integers $k \geq 0$.

We assume that $p_n = \frac{c}{n}$ with $c \in (0, 1)$. From the analysis in Sect. 10.2, we have seen that for $x \in [n]$, the size of the connected component of $G_n(p_n)$ containing x is given by $T = \min\{t \geq 0 : Y_t = 0\}$. (As noted in Sect. 10.2, the quantities T and Y_t depend on x and n, but this dependence is suppressed in the notation.) Let $\hat{Y}_t$ be a random variable distributed according to the distribution $\text{Bin}(n-1, 1-(1-\frac{c}{n})^t)$. Then from Lemma 10.1,

$$P(T > t) \leq P(Y_t > 0) = P(\hat{Y}_t > t - 1) = P(\hat{Y}_t \geq t).$$

(The inequality above is not an equality because we have continued the definition of Y_t past the time T.) Let $\bar{Y}_t$ be a random variable distributed according to the distribution $\text{Bin}(n-1, \frac{tc}{n})$. By Taylor's remainder formula, $(1-x)^t \geq 1 - tx$, for $x \geq 0$ and t a positive integer. Thus, $\frac{tc}{n} \geq 1 - (1-\frac{c}{n})^t$, and consequently $P(\hat{Y}_t \geq t) \leq P(\bar{Y}_t \geq t)$. Thus, we have

$$P(T > t) \leq P(\bar{Y}_t \geq t). \tag{10.16}$$

If $S_{n,t} \sim \text{Bin}(n, \frac{tc}{n})$ as in Proposition 10.1, then $P(\bar{Y}_t \geq t) \leq P(S_{n,t} \geq t)$. Using this with (10.16) and Proposition 10.1, we conclude that there exists a $\kappa > 0$ such that

$$P(T > t) \leq e^{-\kappa t}, \; t \geq 0, \; n \geq 1. \tag{10.17}$$

Let $\gamma > 0$ satisfy $\gamma\kappa > 1$. Then from (10.17) we have

$$P(T > \gamma \log n) \leq e^{-\kappa\gamma \log n} = n^{-\kappa\gamma}. \tag{10.18}$$

We have proven that the probability that the connected component containing x is larger than $\gamma \log n$ is no greater than $n^{-\kappa\gamma}$. There are n vertices in $G_n(p_n)$; thus the probability that at least one of them is in a connected component larger than $\gamma \log n$ is certainly no larger than $n n^{-\kappa\gamma} = n^{1-\kappa\gamma} \to 0$ as $n \to \infty$. This completes the proof of Theorem 10.1. □

10.5 The Galton–Watson Branching Process

We define a random population process in discrete time. Let $\{q_n\}_{n=0}^\infty$ be a nonnegative sequence satisfying $\sum_{n=0}^\infty q_n = 1$. We will refer to $\{q_n\}_{n=0}^\infty$ as the *offspring distribution* of the process. Consider an initial particle alive at time $t = 0$ and set

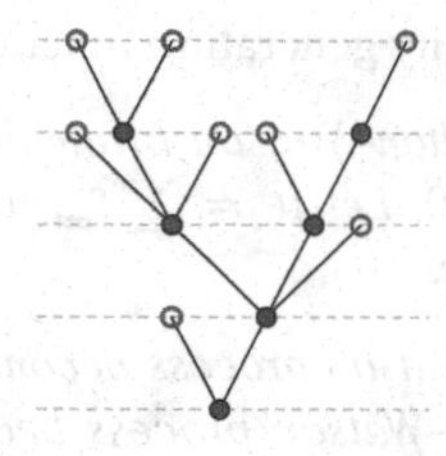

Fig. 10.2 A realization of a branching process that becomes extinct at $n = 5$

$X_0 = 1$ to indicate that the size of the initial population is 1. At time $t = 1$, this particle gives birth to a random number of offspring and then dies. For each $n \in \mathbb{Z}^+$, the probability that there were n offspring is q_n. Let X_1 denote the population size at time 1, namely the number of offspring of the initial particle. In general, at any time $t \geq 1$, all of the X_{t-1} particles alive at time $t-1$ give birth to random numbers of offspring and die. The new number of particles is X_t. The numbers of offspring of the different particles throughout all the generations are assumed independent of one another and are all distributed according to the same offspring distribution $\{q_n\}_{n=0}^{\infty}$.

The random population process $\{X_t\}_{t=0}^{\infty}$ is called a *Galton–Watson branching process.* Clearly, if $X_t = 0$ for some t, then $X_r = 0$ for all $r \geq t$. If this occurs, we say the process becomes *extinct*; otherwise we say that the process *survives*. See Fig. 10.2. If $q_0 = 0$, then the probability of survival is 1. Otherwise, there is a positive probability of extinction, since at any time $t-1$, there is a positive probability (namely $q_0^{X_{t-1}}$) that all of the particles die without leaving any offspring, in which case $X_t = 0$. The most fundamental question we can ask about this process is whether it has a positive probability of surviving.

Let W be a random variable distributed according to the offspring distribution: $P(W = n) = q_n$. Let

$$\mu = EW = \sum_{n=0}^{\infty} nq_n$$

denote the mean number of offspring of a particle. It is easy to show that $EX_{t+1} = \mu EX_t$ (Exercise 10.7), from which it follows that $EX_t = \mu^t$, $t \geq 0$. From this, it follows that if $\mu < 1$, then $\lim_{t\to\infty} EX_t = 0$. Since $EX_t \geq P(X_t \geq 1)$, it follows that $\lim_{t\to\infty} P(X_t \geq 1) = 0$, which means that the process has probability 1 of extinction. The fact that EX_t is growing exponentially in t when $\mu > 1$ would suggest, but not prove, that for $\mu > 1$ the probability of extinction is less than 1. In fact, we can use the method of generating functions to prove the following result. Define

$$\phi(s) = \sum_{n=0}^{\infty} q_n s^n, \ s \in [0, 1]. \tag{10.19}$$

The function $\phi(s)$ is the probability generating function for the distribution $\{q_n\}_{n=0}^{\infty}$.

Theorem 10.3. *Consider a Galton–Watson branching process with offspring distribution $\{q_n\}_{n=0}^{\infty}$, where $q_0 > 0$. Let $\mu = \sum_{n=0}^{\infty} nq_n \in [0,\infty]$ denote the mean number of offspring of a particle.*

(i) If $\mu \leq 1$, then the Galton–Watson process becomes extinct with probability 1.
(ii) If $\mu > 1$, then the Galton–Watson process becomes extinct with probability $\alpha \in (0,1)$, where α is the unique root $s \in (0,1)$ of the equation $\phi(s) = s$.

Proof. If $q_0 + q_1 = 1$, then necessarily, $\mu < 1$. Thus, it follows from the paragraph before the statement of the theorem that extinction occurs with probability 1. Assume now that $q_0 + q_1 < 1$. Since the power series for $\phi(s)$ converges uniformly for $s \in [0, 1-\epsilon]$, for any $\epsilon > 0$, it follows that we can differentiate term by term to get

$$\phi'(s) = \sum_{n=0}^{\infty} nq_n s^{n-1} \geq 0, \ \ \phi''(s) = \sum_{n=0}^{\infty} n(n-1)q_n s^{n-2} \geq 0, \ \ 0 \leq s < 1.$$

In particular then, since $q_0 + q_1 < 1$, ϕ is a strictly convex function on $[0,1]$, and consequently, so is $\psi(s) := \phi(s) - s$. We have $\psi(0) = q_0 > 0$ and $\psi(1) = 0$. Also, $\lim_{s\to 1^-} \psi'(s) = \lim_{s\to 1^-} \phi'(s) - 1 = \mu - 1$. Since ψ is strictly convex, it follows that if $\mu \leq 1$, then $\psi'(s) < 0$ for $s \in [0,1)$, and consequently $\psi(s) > 0$, for $s \in [0,1)$. However, if $\mu > 1$, then $\psi'(s) > 0$ for $s < 1$ and sufficiently close to 1. Using this along with the strict convexity and the fact that $\psi(0) > 0$ and $\psi(1) = 0$, it follows that there exists a unique $\alpha \in (0,1)$ such that $\psi(\alpha) = 0$ and that $\psi(s) > 0$, for $s \in (0,\alpha)$, and $\psi(s) < 0$, for $s \in (\alpha, 1)$. (The reader should verify this.) We have thus shown that

$$\begin{aligned}
&\text{the smallest root } \alpha \in [0,1] \text{ of the equation } \phi(z) = z \text{ satisfies}\\
&\alpha \in (0,1), \text{ if } \mu > 1, \text{ and } \alpha = 1, \text{ if } \mu \leq 1. \text{ Furthermore, in the case } \mu > 1,\\
&\text{one has } \phi(s) > s, \text{ for } s \in [0,\alpha), \text{ and } \phi(s) < s, \text{ for } s \in (\alpha, 1).
\end{aligned} \tag{10.20}$$

Now let $\kappa_t := P(X_t = 0)$ denote the probability that extinction has occurred by time t. Of course, $\kappa_0 = 0$. We claim that

$$\kappa_t = \phi(\kappa_{t-1}), \text{ for } t \geq 1. \tag{10.21}$$

To prove this, first note that when $t = 1$, (10.21) says that $\kappa_1 = \phi(0) = q_0$, which is of course true. Now consider $t > 1$. We first calculate $P(X_t = 0|X_1 = n)$, the probability that $X_t = 0$, conditioned on $X_1 = n$. By the conditioning, at time $t = 1$, there are n particles, and each of these particles will contribute independently to the population size X_t at time t, through $t-1$ generations of branching. In order to have $X_t = 0$, each of these n "new" branching processes must become extinct by time $t-1$. The probability that any one of them becomes extinct by time $t-1$ is, by definition, κ_{t-1}. By the independence, it follows that the probability that they all become extinct by time $t-1$ is κ_{t-1}^n. We have thus proven that

$$P(X_t = 0|X_1 = n) = \kappa_{t-1}^n.$$

Since $P(X_1 = n) = q_n$, we conclude that

$$\kappa_t = P(X_t = 0) = \sum_{n=0}^{\infty} P(X_1 = n)P(X_t = 0|X_1 = n) = \sum_{n=0}^{\infty} q_n \kappa_{t-1}^n = \phi(\kappa_{t-1}),$$

proving (10.21). From its definition, κ_t is nondecreasing, and $\kappa_{\text{ext}} := \lim_{t\to\infty} \kappa_t$ is the extinction probability. Letting $t \to \infty$ in (10.21) gives

$$\kappa_{\text{ext}} = \phi(\kappa_{\text{ext}}). \tag{10.22}$$

It follows immediately from (10.22) and (10.20) that $\kappa_{\text{ext}} = 1$, if $\mu \leq 1$. If $\mu > 1$, then there are two roots $s \in [0, 1]$ of the equation $\phi(s) = s$, namely $s = \alpha$ and $s = 1$. If $\kappa_{\text{ext}} = 1$, then $\kappa_t > \alpha$ for sufficiently large t, and then by (10.20) and (10.21), for such t, we have $\kappa_{t+1} = \phi(\kappa_t) < \kappa_t$, which contradicts the fact that κ_t is nondecreasing. Thus, we conclude that $\kappa_{\text{ext}} = \alpha$. □

At one point in the proof of Theorem 10.2, we will use the above result on the extinction probability of a Galton–Watson branching process. However, we will need to consider this process in an alternative form. In the original formulation, at time t, the entire population of size X_{t-1} that was alive at time $t-1$ reproduces and dies, and then X_t is the new population size. In other words, time t referred to the tth generation of particles. In our alternative formulation, at each time, t only *one* of the particles that was alive at time $t-1$ reproduces and dies. Thus, as before, we have $X_0 = 1$ to denote that we start with a single particle, and X_1 denotes the number of offspring that the original particle produces before it dies. At time $t = 2$, instead of having all X_1 particles reproduce and die simultaneously, we choose (arbitrarily) just one of these particles that was alive at time $t = 1$ and have it reproduce and die. Then X_2 is equal to the new total population. We continue in this way, at each step choosing just one of the particles that was alive at the previous step. Since in any case, the number of offspring of any particle is independent of the number of offspring of the other particles, it is clear that this new process has the same extinction probability as the original one.

10.6 Proof of Theorem 10.2

We assume that $p_n = \frac{c}{n}$, with $c > 1$. From the analysis in Sect. 10.2, we have seen that for $x \in [n]$, the size of the connected component of $G_n(p_n)$ containing x is given by $T = \min\{t \geq 0 : Y_t = 0\}$.

Consider a Galton–Watson branching process $\{X_t\}_{t=0}^{\infty}$ in the alternative form described at the end of Sect. 10.5, and let the offspring distribution be the Poisson distribution with parameter c; that is, $q_m = e^{-c}\frac{c^m}{m!}$. The probability generating function of this distribution is given by

$$\phi(s) = \sum_{m=0}^{\infty} q_m s^m = \sum_{m=0}^{\infty} e^{-c}\frac{c^m}{m!}s^m = e^{c(s-1)}. \tag{10.23}$$

The expected number of offspring is equal to c. Since $c > 1$, it follows from Theorem 10.3 that the extinction time $T_{\text{ext}} = \inf\{t \geq 1 : X_t = 0\}$ satisfies

$$P(T_{\text{ext}} < \infty) = \alpha, \tag{10.24}$$

where $\alpha \in (0, 1)$ is the unique solution $s \in (0, 1)$ to the equation $\phi(s) = s$, that is, to the equation $e^{c(s-1)} = s$. Substituting $z = 1 - s$ in this equation, this becomes

$$1 - \alpha \text{ is the unique root } z \in (0,1) \text{ of the equation } 1 - e^{-cz} - z = 0. \tag{10.25}$$

Let $\{W_t\}_{t=1}^{\infty}$ be a sequence of independent, identically distributed random variables distributed according to the Poisson distribution with parameter c. If $X_{t-1} \neq 0$, then W_t will serve as the number of offspring of the particle chosen for reproduction and death from among the X_{t-1} particles alive at time $t-1$. Then we may represent the process $\{X_t\}_{t=0}^{\infty}$ by $X_0 = 1$ and

$$X_t = X_{t-1} + W_t - 1,\ 1 \leq t \leq T_{\text{ext}}. \tag{10.26}$$

If $T_{\text{ext}} < \infty$, then of course $X_t = 0$ for all $t \geq T_{\text{ext}}$. For any fixed $t \geq 1$, as soon as one knows the values of $\{W_s\}_{s=1}^{t}$, one knows the values of $\{X_s\}_{s=1}^{t}$. (We note that it might happen that these values of $\{W_s\}_{s=1}^{t}$ result in $X_{s_0} = 0$ for some $s_0 < t$, in which case the values of $\{W_s\}_{s=s_0+1}^{t}$ are superfluous for determining the values of $\{X_s\}_{s=1}^{t}$.) If $\bar{r} := \{r_s\}_{s=1}^{t}$ are the values obtained for $\{W_s\}_{s=1}^{t}$, let $\bar{l} := \{l_s\}_{s=1}^{t}$ denote the corresponding values for $\{X_s\}_{s=1}^{t}$. We write $\bar{l} = \bar{l}(\bar{r})$. Note that $T_{\text{ext}} \geq t$ occurs if and only if $l_s > 0$, for $0 \leq s \leq t-1$, or, equivalently, if and only if $l_{t-1} > 0$.

Now consider the process $\{Y_t\}_{t=0}^{\infty}$ introduced in Sect. 10.2. Recall that T is equal to the smallest t for which $Y_t = 0$. Note from (10.3) and (10.26) that $\{Y_t\}_{t=0}^{T}$ is defined recursively in a way very similar to the way $\{X_t\}_{t=0}^{T_{\text{ext}}}$ is defined. The difference is that the independent sequence of random variables $\{W_t\}_{t=1}^{\infty}$

distributed according to the Poisson distribution with parameter c is replaced by the sequence $\{Z_t\}_{t=1}^{\infty}$. The distribution of these latter random variables is given by (10.4) and (10.5). (As noted in Sect. 10.2, Y_t, T, and Z_t depend on x and n, but that dependence has been suppressed in the notation.) Because the form of the recursion formula is the same for $\{X_s\}_{s=1}^{T_{\text{ext}}}$ and $\{Y_s\}_{s=1}^{T}$, and because $X_0 = Y_0 = 1$, it follows that if $\bar{r} = \{r_s\}_{s=1}^{t}$ are the values obtained for $\{Z_s\}_{s=1}^{t}$, and if $\bar{l}(\bar{r})$ satisfies $l_{t-1} > 0$, then $\bar{l}(\bar{r})$ are the corresponding values for $\{Y_s\}_{s=1}^{t}$.

Since the random variables $\{W_s\}_{s=1}^{t}$ are independent, we have

$$P(\{W_s\}_{s=1}^{t} = \bar{r}) = \prod_{s=1}^{t} P(W_s = r_s). \tag{10.27}$$

By (10.4) and by the conditional independence condition (10.5), if $l_{t-1} > 0$, we have

$$P(\{Z_s\}_{s=1}^{t} = \bar{r}) = \prod_{s=1}^{t} P(Z_s = r_s | Y_{s-1} = l_{s-1}), \tag{10.28}$$

where, for convenience, we define $l_0 = 1$. By (10.4), the distribution of Z_s, conditioned on $Y_{s-1} = l_{s-1}$, is given by $\text{Bin}(n - s - l_{s-1} + 1, \frac{c}{n})$. Since $\lim_{n\to\infty} (n - s - l_{s-1} + 1)\frac{c}{n} = c$, it follows from the Poisson approximation to the binomial distribution (see Proposition A.3 in Appendix A) that

$$\lim_{n\to\infty} P(Z_s = r_s | Y_{s-1} = l_{s-1}) = P(W_s = r_s).$$

Thus, we conclude from (10.27) and (10.28) that for any fixed t,

$$\lim_{n\to\infty} P(\{Z_s\}_{s=1}^{t} = \bar{r}) = P(\{W_s\}_{s=1}^{t} = \bar{r}),$$
$$\text{for all } \bar{r} = \{r_s\}_{s=1}^{t} \text{ for which } l_{t-1}(\bar{r}) > 0,$$

and, consequently, that for any fixed t,

$$\lim_{n\to\infty} P(\{Y_s\}_{s=1}^{t} = \bar{l}) = P(\{X_s\}_{s=1}^{t} = \bar{l}), \text{ for all } \bar{l} = \{l_s\}_{s=1}^{t} \text{ satisfying } l_{t-1} > 0. \tag{10.29}$$

Since T_{ext}, the extinction time for $\{X_t\}_{t=0}^{\infty}$, is the smallest t for which $X_t = 0$, and $T_{\text{ext}} \geq t$ is equivalent to $l_{t-1} > 0$, and since T is the smallest t for which $Y_t = 0$, it follows from (10.29) that

$$\lim_{n\to\infty} P(T \leq t) = P(T_{\text{ext}} \leq t), \text{ for any fixed } t \geq 1. \tag{10.30}$$

From (10.24), we have $\lim_{t\to\infty} P(T_{\text{ext}} \le t) = \alpha$; thus, for any $\epsilon > 0$, there exists an integer τ_ϵ such that $P(T_{\text{ext}} \le t) \in (\alpha - \frac{\epsilon}{2}, \alpha)$, if $t \ge \tau_\epsilon$. It then follows from (10.30) that there exists an $n_{1,\epsilon} = n_{1,\epsilon}(t)$ such that

$$P(T \le t) \in (\alpha - \epsilon, \alpha + \epsilon), \text{ if } t \ge \tau_\epsilon \text{ and } n \ge n_{1,\epsilon}(t). \tag{10.31}$$

We now analyze the probabilities $P(\epsilon n \le T \le (1-\alpha-\epsilon)n)$ and $P((1-\alpha+\epsilon)n \le T \le n)$. We will show that these probabilities are very small. From (10.25), it follows that $1 - e^{-cz} > z$, for $z \in (0, 1-\alpha)$, and $1 - e^{-cz} < z$, for $z \in (1-\alpha, 1]$. Consequently, for $\epsilon > 0$, choosing $\delta = \delta(\epsilon)$ sufficiently small, we have

$$1 - e^{-cz} - 2\delta > z, \text{ for } z \in (\epsilon, 1-\alpha-\epsilon);\ 1 - e^{-cz} + 2\delta < z, \text{ for } z \in (1-\alpha+\epsilon, 1]. \tag{10.32}$$

Since T is the smallest t for which $Y_t = 0$, we have

$$\{\epsilon n \le T \le (1-\alpha-\epsilon)n\} \subset \cup_{\epsilon n \le t \le (1-\alpha-\epsilon)n}\{Y_t \le 0\}.$$

(Recall that Y_t has also been defined recursively for $t > T$ and can take on negative values for such t.) Thus, letting $\hat{Y}_t$ be the random variable distributed according to the distribution $\text{Bin}(n-1, 1-(1-\frac{c}{n})^t)$, it follows from Lemma 10.1 that

$$P(\epsilon n \le T \le (1-\alpha-\epsilon)n) \le \sum_{\epsilon n \le t \le (1-\alpha-\epsilon)n} P(\hat{Y}_t \le t-1). \tag{10.33}$$

One has $\lim_{n\to\infty}(1-\frac{c}{n})^{bn} = e^{-cb}$, uniformly over b in a bounded set. (The reader should verify this by taking the logarithm of $(1-\frac{c}{n})^{bn}$ and applying Taylor's formula.) Applying this with $b = \frac{t}{n}$, with $0 \le t \le n$, it follows that $(1-\frac{c}{n})^t - e^{-\frac{ct}{n}}$ is small for large n, uniformly over $t \in [0, n]$. Thus, for $\delta = \delta(\epsilon)$, which has been defined above, there exists an $n_{2,\delta} = n_{2,\delta(\epsilon)}$ such that $1-(1-\frac{c}{n})^t \ge 1 - e^{-\frac{ct}{n}} - \delta$, for $n \ge n_{2,\delta}$ and $0 \le t \le n$. Let $\bar{Y}_t$ be a random variable distributed according to the distribution $\text{Bin}(n-1, 1-e^{-\frac{ct}{n}} - \delta)$. Then $P(\hat{Y}_t \le t-1) \le P(\bar{Y}_t \le t-1)$, if $n \ge n_{2,\delta}$. Using this with (10.33), we obtain

$$P(\epsilon n \le T \le (1-\alpha-\epsilon)n) \le \sum_{\epsilon n \le t \le (1-\alpha-\epsilon)n} P(\bar{Y}_t \le t-1),\ n \ge n_{2,\delta}. \tag{10.34}$$

Every t in the summation on the right hand side of (10.34) is of the form $t = b_n n$, with $\epsilon \le b_n \le 1-\alpha-\epsilon$. Thus, it follows from (10.32) that $1 - e^{-\frac{ct}{n}} - \delta = 1 - e^{-cb_n} - \delta \ge b_n + \delta$. We now apply part (ii) of Proposition 10.2 with $n-1$ in place of n, with $\rho = 1 - e^{-cb_n} - \delta$, and with $\rho_0 = b_n$. Note that ρ and ρ_0 are bounded from 0 and from 1 as n varies and as t varies over the above range. Also, we have $\rho > \rho_0 + \delta$. Consequently, there exists a constant $\kappa > 0$ such that $\kappa(\rho_0, \rho) \ge \kappa$, for all ρ, ρ_0 as above. Thus, we have for $n \ge n_{2,\delta}$,

$$P(\bar{Y}_t \le t-1) = P(\bar{Y}_t \le b_n n - 1) \le P(\bar{Y}_t \le b_n(n-1)) \le e^{-\kappa(n-1)}. \quad (10.35)$$

From (10.34) and (10.35) we conclude that

$$P(\epsilon n \le T \le (1-\alpha-\epsilon)n) \le (1-\alpha)ne^{-\kappa(n-1)},\ n \ge n_{2,\delta(\epsilon)}. \quad (10.36)$$

A very similar analysis shows that

$$P((1-\alpha+\epsilon)n \le T \le n) \le \alpha n e^{-\kappa(n-1)},\ n \ge n_{3,\delta(\epsilon)}, \quad (10.37)$$

for some $n_{3,\delta} = n_{3,\delta(\epsilon)}$. This is left to the reader as Exercise 10.8.

We now analyze the probability $P(t < T < \epsilon n)$, for fixed t. As in (10.33), we have

$$P(t < T < \epsilon n) \le \sum_{t<s<\epsilon n} P(\hat{Y}_s \le s-1), \quad (10.38)$$

where, we recall, $\hat{Y}_s$ is distributed according to the distribution $\text{Bin}(n-1, 1-(1-\frac{c}{n})^s)$. Let $\tilde{Y}_s$ be a random variable distributed according to the distribution $\text{Bin}(n-1, (1-\frac{c}{n})^s)$. Then

$$P(\hat{Y}_s \le s-1) = P(\tilde{Y}_s \ge n-s). \quad (10.39)$$

As in the proofs of Propositions 10.1 and 10.2, we have for any $\lambda > 0$,

$$P(\tilde{Y}_s \ge n-s) \le e^{-\lambda(n-s)} Ee^{\lambda \tilde{Y}_s}. \quad (10.40)$$

We can represent the random variable $\tilde{Y}_s$ as $\tilde{Y}_s = \sum_{j=1}^{n-1} B_j$, where the $\{B_j\}_{j=1}^{n-1}$ are independent and identically distributed Bernoulli random variables with parameter $(1-\frac{c}{n})^s$; that is, $P(B_j = 1) = 1 - P(B_j = 0) = (1-\frac{c}{n})^s$. Using the fact that these random variables are independent and identically distributed, we have

$$Ee^{\lambda \tilde{Y}_s} = \prod_{j=1}^{n-1} Ee^{\lambda B_j} = \left[(1-\frac{c}{n})^s)e^{\lambda} + 1 - (1-\frac{c}{n})^s\right]^{n-1}. \quad (10.41)$$

Thus, from (10.40) and (10.41), we obtain

$$P(\tilde{Y}_s \ge n-s) \le e^{-\lambda(n-s)}\left[(1-\frac{c}{n})^s e^{\lambda} + 1 - (1-\frac{c}{n})^s\right]^{n-1}. \quad (10.42)$$

We now substitute $n = Ms$ in (10.42) to obtain

$$P(\tilde{Y}_s \geq n - s) \leq e^{-\lambda s(M-1)}\big[(1 - \frac{c}{Ms})^s(e^\lambda - 1) + 1\big]^{Ms-1} \leq$$
$$e^{-\lambda s(M-1)}\big[(1 - \frac{c}{Ms})^s(e^\lambda - 1) + 1\big]^{Ms} = e^{s\left[-\lambda(M-1)+M\log\left((1-\frac{c}{Ms})^s(e^\lambda-1)+1\right)\right]},$$
$$\text{for all } \lambda > 0. \tag{10.43}$$

We will show that for an appropriate choice of $\lambda > 0$, the expression in the square brackets above is negative and bounded away from 0 for all $s \geq 1$ and sufficiently large M. Let

$$f_{s,M}(\lambda) := -\lambda(M-1) + M\log\big((1 - \frac{c}{Ms})^s(e^\lambda - 1) + 1\big).$$

Then $f_{s,M}(0) = 0$ and

$$f'_{s,M}(\lambda) = -(M-1) + \frac{M(1-\frac{c}{Ms})^s e^\lambda}{(1-\frac{c}{Ms})^s(e^\lambda - 1) + 1}. \tag{10.44}$$

For any fixed λ, defining $g(y) = \frac{ye^\lambda}{y(e^\lambda-1)+1}$, for $y > 0$, it is easy to check that $g'(y) > 0$; therefore, g is increasing. The last term on the right hand side of (10.44) is $Mg(y)$, with $y = (1 - \frac{c}{Ms})^s$. Since $1 - x \leq e^{-x}$, for $x \geq 0$, we have $(1 - \frac{c}{Ms})^s \leq e^{-\frac{c}{M}}$, if $n = Ms \geq c$, and thus the last term on the right hand side of (10.44) is bounded from above by $Mg(e^{-\frac{c}{M}})$, independent of s, for $s \geq \frac{c}{M}$. Thus, from (10.44), we have

$$f'_{s,M}(\lambda) \leq -(M-1) + \frac{Me^{-\frac{c}{M}}e^\lambda}{e^{-\frac{c}{M}}(e^\lambda - 1) + 1} = -M + 1 + \frac{Me^\lambda}{e^\lambda - 1 + e^{\frac{c}{M}}} =$$
$$1 + M\frac{1 - e^{\frac{c}{M}}}{e^\lambda - 1 + e^{\frac{c}{M}}}, \text{ for all } s \geq \frac{c}{M}.$$

Since $\lim_{M\to\infty} M\frac{1-e^{\frac{c}{M}}}{e^\lambda-1+e^{\frac{c}{M}}} = -ce^{-\lambda}$, uniformly over $\lambda \in [0,1]$, and since $c > 1$, it follows that there exists a $\lambda_0 > 0$ and an M_0 such that if $\lambda \in [0, \lambda_0]$ and $M \geq M_0$, then $f'_{s,M}(\lambda) \leq \frac{1-c}{2}$, for all $s \geq 1$. It then follows that $f_{s,M}(\lambda_0) \leq -\frac{\lambda_0(c-1)}{2}$, for all $M \geq M_0$ and $s \geq 1$. Choosing $\lambda = \lambda_0$ in (10.43) and using this last inequality for $f_{s,M}(\lambda_0)$, we conclude that

$$P(\tilde{Y}_s \geq n - s) \leq e^{-\frac{\lambda_0(c-1)}{2}s}, \text{ for } n \geq M_0 s,\ s \geq 1. \tag{10.45}$$

From (10.38), (10.39), and (10.45), we obtain the estimate

$$P(t < T < \epsilon n) \le \sum_{t<s<\epsilon n} e^{-\frac{\lambda_0(c-1)}{2}s} < \sum_{s=t}^{\infty} e^{-\frac{\lambda_0(c-1)}{2}s} = \frac{e^{-\frac{\lambda_0(c-1)}{2}t}}{1-e^{-\frac{\lambda_0(c-1)}{2}}},$$

$$\text{if } \epsilon \le \frac{1}{M_0}. \tag{10.46}$$

Now (10.31), (10.36), (10.37), and (10.46) guarantee that for any $\epsilon \in (0,1)$, we can choose t_ϵ and n_ϵ such that for all $n \ge n_\epsilon$, one has

$$\begin{aligned} &\alpha - \epsilon \le P(T \le t_\epsilon) \le \alpha + \epsilon; \\ &P\big(T > t_\epsilon, T \notin ((1-\alpha-\epsilon)n, (1-\alpha+\epsilon)n)\big) \le \epsilon; \\ &1-\alpha-2\epsilon \le P\big(T \in ((1-\alpha-\epsilon)n, (1-\alpha+\epsilon)n)\big) \le 1-\alpha+\epsilon, \\ &\text{for all } n \ge n_\epsilon. \end{aligned} \tag{10.47}$$

(The third set of inequalities above is a consequence of the first two sets of inequalities.)

We recall that the above estimates have been obtained when $p = \frac{c}{n}$, with $c > 1$, and where $1-\alpha = 1-\alpha(c)$ is the unique root $z \in (0,1)$ of the equation $1-e^{-cz}-z=0$. The reader can check that the above estimates hold uniformly for $c \in [c_1, c_2]$, for any $1 < c_1 < c_2$. Thus, consider as before a fixed $c > 1$ and $\alpha = \alpha(c)$, and let $\delta > 0$ satisfy $c - \delta > 1$. For $c' \in [c-\delta, c]$, let $\alpha' := \alpha(c')$. Then for all $\epsilon > 0$, there exists a $t_\epsilon > 0$ and a $n_\epsilon > 0$ such that for all $n \ge n_\epsilon$ and all $c' \in [c-\delta, c]$, one has for the graph $G(n, \frac{c'}{n})$,

$$\begin{aligned} &\alpha' - \epsilon \le P(T \le t_\epsilon) \le \alpha' + \epsilon; \\ &P\big(T > t_\epsilon, T \notin ((1-\alpha'-\epsilon)n, (1-\alpha'+\epsilon)n)\big) \le \epsilon; \\ &1-\alpha'-2\epsilon \le P\big(T \in ((1-\alpha'-\epsilon)n, (1-\alpha'+\epsilon)n)\big) \le 1-\alpha'+\epsilon, \\ &\text{for all } n \ge n_\epsilon. \end{aligned} \tag{10.48}$$

Return now to our graph $G(n, \frac{c}{n})$, with n considerably larger than the n_ϵ in (10.48). (We will quantify "considerably larger" a bit later on.) Recall that we started out by choosing arbitrarily some vertex x in the graph $G(n, \frac{c}{n})$, and then applied our algorithm, obtaining T, which is the size of the connected component containing x. Call this the first step in a "game." If it results in $T \le t_\epsilon$, say that a "draw" occurred on the first step. If it results in $(1-\alpha-\epsilon)n < T < (1-\alpha+\epsilon)n$, say that a "win" occurred on the first step. Otherwise, say that a "loss" occurred on the first step. If a win or a loss occurs on this first step, we stop the procedure and say that the game ended in a win or loss, respectively. If a draw occurs, then consider the remaining $n - T$ vertices that are not in the connected component containing x, and consider the corresponding edges. This gives a graph of size $n' = n - T$. Note that by the definition of the algorithm, there is no pair of points in this new graph that has already been checked by the algorithm. Therefore, the conditional edge probabilities for this new graph, conditioned on having implemented the algorithm, are as before, namely $\frac{c}{n}$, independently for each edge. This edge probability can be written as $p_{n'} = \frac{c'}{n'}$, where $c' = \frac{n-T}{n}c$. Now $T \le t_\epsilon$. Thus, if $n \ge n_\epsilon$ is sufficiently large, then $c' \in [c - \delta, c]$ and $n' = n - T \ge n_\epsilon$, so the estimates (10.48) (with n replaced by n') will hold for this new graph, which has n' vertices and edge probabilities $p_{n'} = \frac{c'}{n'}$. Choose an arbitrary vertex x_1 from this new graph and repeat the above algorithm on the new graph. Let T_1 denote the random variable T for this second step. If a win or a loss occurs on the second step of the game, then we stop the game and say that the game ended in a win or a loss, respectively. (Of course, here we define win, loss, and draw in terms of T_1, n', and α' instead of T, n, and α. However, the same t_ϵ is used.) If a draw occurs on this second step, then we consider the $n' - T_1 = n - T - T_1$ vertices that are neither in the connected component of x nor of x_1. We continue like this for a maximum of M_ϵ steps, where M_ϵ is chosen sufficiently large to satisfy $\big(\alpha(c - \delta) + \epsilon\big)^{M_\epsilon} < \epsilon$. (We work with $\epsilon > 0$ sufficiently small so that $\alpha(c - \delta) + \epsilon < 1$.) The reason for this choice of M_ϵ will become clear below. If after M_ϵ steps, a win or a loss has not occurred, then we declare that the game has ended in a draw. Note that the smallest possible graph size that can ever be used in this game is $n - t_\epsilon(M_\epsilon - 1)$. The smallest modified value of c that can ever be used is $\frac{n - t_\epsilon(M_\epsilon - 1)}{n}c$. We can now quantify what we meant when we said at the outset of this paragraph that we are choosing n "considerably larger" than n_ϵ. We choose n sufficiently large so that $n - t_\epsilon(M_\epsilon - 1) \ge n_\epsilon$ and so that $\frac{n - t_\epsilon(M_\epsilon - 1)}{n}c \ge c - \delta$. Thus, the estimates in (10.48) are valid for all of the steps of the game.

It is easy to check that $\alpha = \alpha(c)$ is decreasing for $c > 1$. Thus, if the game ends in a win, then there is a connected component of size between $(1 - \alpha(c - \delta) - \epsilon)n$ and $(1 - \alpha(c) + \epsilon)n$. What is the probability that the game ends in a win? Let W denote the event that the game ends in a win, let D denote the event that it ends in a draw, and let L denote the event that it ends in a loss. We have

$$P(W) = 1 - P(L) - P(D). \tag{10.49}$$

The game ends in a draw if there was a draw on M_ϵ consecutive steps. Since on any given step the probability of a draw is no greater than $\alpha(c - \delta) + \epsilon$, the probability of obtaining M_ϵ consecutive draws is no greater than $\big(\alpha(c - \delta) + \epsilon\big)^{M_\epsilon}$; so by the choice of M_ϵ, we have

$$P(D) \le \big(\alpha(c - \delta) + \epsilon\big)^{M_\epsilon} < \epsilon. \tag{10.50}$$

Let D^c denote the complement of D; that is, $D^c = W \cup L$. Obviously, we have $L = L \cap D^c$. Then we have

$$P(L) = P(L \cap D^c) = P(D^c)P(L|D^c) \le P(L|D^c). \tag{10.51}$$

If one played a game with three possible outcomes on each step—win, loss, or draw—with respective nonzero probabilities p', q', and r', and the outcomes of all the steps were independent of one another, and one continued to play step after step until either a win or a loss occurred, then the probability of a win would be $\frac{p'}{p'+q'}$ and the probability of a loss would be $\frac{q'}{p'+q'}$ (Exercise 10.9). Conditioned on D^c, our game essentially reduces to this game. However, the probabilities of win and loss and draw are not exactly fixed, but can vary a little according to (10.48). Thus, we can conclude that

$$P(L|D^c) \le \frac{\epsilon}{1-\alpha(c-\delta)-2\epsilon+\epsilon} = \frac{\epsilon}{1-\alpha(c-\delta)-\epsilon}. \tag{10.52}$$

From (10.49)–(10.52) we obtain

$$P(W) \ge 1-\epsilon-\frac{\epsilon}{1-\alpha(c-\delta)-\epsilon}. \tag{10.53}$$

In conclusion, we have demonstrated the following. Consider any $c>1$ and any $\delta>0$ such that $c-\delta>1$. Then for each sufficiently small $\epsilon>0$ and sufficiently large n depending on ϵ, with probability at least $1-\epsilon-\frac{\epsilon}{1-\alpha(c-\delta)-\epsilon}$ there will exist a connected component of $G(n,\frac{c}{n})$ of size between $(1-\alpha(c-\delta)-\epsilon)n$ and $(1-\alpha(c)+\epsilon)n$. If the connected component above, which has been shown to exist with probability close to 1 and which is of size around $(1-\alpha)n$, is in fact with probability close to 1 the largest connected component, then the above estimates prove (10.1), since by (10.25) the β defined in the statement of the theorem is in fact $1-\alpha$. Thus, to complete the proof of (10.1) and (10.2), it suffices to prove that with probability approaching 1 as $n\to\infty$, every other component of $G(n,\frac{c}{n})$ is of size $O(\log n)$, as $n\to\infty$. In fact, we will prove here the weaker result that with probability approaching 1 as $n\to\infty$, every other component is of size $o(n)$ as $n\to\infty$. In Exercise 10.10, the reader is guided through a proof that every other component is of size $O(\log n)$.

To prove that every other component is of size $o(n)$ with probability approaching 1 as $n\to\infty$, assume to the contrary. Then for an unbounded sequence of n's, the following holds. As above, with probability at least $1-\epsilon-\frac{\epsilon}{1-\alpha(c-\delta)-\epsilon}$, there will exist a connected component of $G(n,\frac{c}{n})$ of size between $(1-\alpha(c-\delta)-\epsilon)n$ and $(1-\alpha(c)+\epsilon)n$, and by our assumption, for some $\gamma>0$, with probability at least γ, there will be another connected component of size at least γn. We may take $\gamma<1-\alpha(c-\delta)-\epsilon$. But if this were true, then at the first step of our algorithm, when we randomly selected a vertex x, the probability that it would be in a connected component of size at least γn would be at least

$$\Big(1-\epsilon-\frac{\epsilon}{1-\alpha(c-\delta)-\epsilon}\Big)\frac{(1-\alpha(c-\delta)-\epsilon)n}{n}+\gamma\frac{\gamma n}{n}.$$

For ϵ and δ sufficiently small, this number will be larger than $1-\alpha(c)+\frac{\gamma^2}{2}$, in which case the algorithm would have to give $P(T\le t_\epsilon)<\alpha(c)-\frac{\gamma^2}{2}$. However, for $\epsilon>0$ sufficiently small, this contradicts the first line of (10.47). □

Exercise 10.1. This exercise refers to Remark 3 after Theorem 10.2. Prove that for any $\epsilon>0$ and large n, the number of edges of $G_n(\frac{c}{n})$ is equal to $\frac{1}{2}cn+O(n^{\frac{1}{2}+\epsilon})$ with high probability. Show directly that $\beta(c)\le\frac{c}{2}$, for $1<c<2$, where $\beta(c)$ is as in Theorem 10.2.

Exercise 10.2. Let D_n denote the number of disconnected vertices in the Erdős–Rényi graph $G_n(p_n)$. For this exercise, it will be convenient to represent D_n as a sum of indicator random variables. Let $D_{n,i}$ be equal to 1 if the vertex i is disconnected and equal to 0 otherwise. Then $D_n=\sum_{i=1}^n D_{n,i}$.

(a) Calculate ED_n.
(b) Calculate ED_n^2. (Hint: Write $ED_n^2=E(\sum_{i=1}^n D_{n,i})(\sum_{j=1}^n D_{n,j})$.)

Exercise 10.3. In this exercise, you are guided through a proof of the result noted in Remark 2 after Theorem 10.2, namely that:
if $p_n=\frac{\log n+c_n}{n}$, then as $n\to\infty$, the probability that the Erdős–Rényi graph $G_n(p_n)$ possesses at least one disconnected vertex approaches 0 if $\lim_{n\to\infty}c_n=\infty$, while for any M, the probability that it possesses at least M disconnected vertices approaches 1 if $\lim_{n\to\infty}c_n=-\infty$.
Let D_n be as in Exercise 10.2, with $p_n=\frac{\log n+c_n}{n}$.

(a) Use Exercise 10.2(a) to show that $\lim_{n\to\infty}ED_n$ equals 0 if $\lim_{n\to\infty}c_n=\infty$ and equals ∞ if $\lim_{n\to\infty}c_n=-\infty$. (Hint: Consider $\log ED_n$ and note that by Taylor's remainder theorem, $\log(1-x)=-x-\frac{1}{(1-x^*)^2}\frac{x^2}{2}$, for $0<x<1$, where $x^*=x^*(x)$ satisfies $0<x^*<x$.)
(b) Use (a) to show that if $\lim_{n\to\infty}c_n=\infty$, then $\lim_{n\to\infty}P(D_n=0)=1$.
(c) Use Exercise 10.2(b) to calculate ED_n^2.
(d) Show that if $\lim_{n\to\infty}c_n=-\infty$, then the variance $\sigma^2(D_n)$ satisfies $\sigma^2(D_n)=o\big((ED_n)^2\big)$. (Hint: Recall that $\sigma^2(D_n)=ED_n^2-(ED_n)^2$.)
(e) Use Chebyshev's inequality with (a) and (d) to conclude that if $\lim_{n\to\infty}c_n=-\infty$, then for any M, $\lim_{n\to\infty}P(D_n\ge M)=1$.

Exercise 10.4. Recall from Chap. 5 that the probability generating function $P_X(s)$ of a nonnegative random variable X taking integral values is defined by

$$P_X(s)=Es^X=\sum_{i=0}^{\infty}s^iP(X=i).$$

The probability generating function of a random variable X uniquely characterizes its distribution, because $\frac{P_X^{(i)}(0)}{i!}=P(X=i)$.

(a) Let $X \sim \text{Bin}(n, p)$. Show that $P_X(s) = (ps + 1 - p)^n$.

(b) Let $Z \sim \text{Bin}(n, p)$, and let $Y \sim \text{Bin}(Z, p')$, by which is meant that conditioned on $Z = m$, the random variable Y is distributed according to $\text{Bin}(m, p')$. Calculate $P_Y(s)$ by writing

$$P_Y(s) = Es^Y = \sum_{m=0}^{n} E(s^Y|Z = m)P(Z = m),$$

and conclude that $Y \sim \text{Bin}(n, pp')$. Conclude from this that (10.7) and (10.9) imply (10.6).

Exercise 10.5. Let $f(\lambda) = e^{-\lambda\rho_0}(\rho e^\lambda + 1 - \rho)$, with $0 < \rho < \rho_0 < 1$. Show that $\inf_{\lambda \geq 0} f(\lambda)$ is attained at some $\lambda_0 > 0$ and that $f(\lambda_0) = \left(\frac{1-\rho}{1-\rho_0}\right)^{1-\rho_0}\left(\frac{\rho}{\rho_0}\right)^{\rho_0} \in (0, 1)$.

Exercise 10.6. If $X \sim \text{Bin}(n, p)$, then X can be represented as $X = \sum_{i=1}^n B_i$, where $\{B_i\}_{i=1}^n$ are independent and identically distributed random variables distributed according to the Bernoulli distribution with parameter p; that is, $P(B_i = 1) = 1 - P(B_i = 0) = p$.

(a) Use the above representation to prove that

$$\begin{aligned} &\text{if } X_i \sim \text{Bin}(n_i, p),\ i = 1, 2, \text{ and } n_1 > n_2, \text{ then} \\ &P(X_1 \geq k) \geq P(X_2 \geq k), \text{ for all integers } k \geq 0, \end{aligned} \tag{10.54}$$

and that

$$\begin{aligned} &\text{if } X_i \sim \text{Bin}(n, p_i),\ i = 1, 2, \text{ and } p_1 > p_2, \text{ then} \\ &P(X_1 \geq k) \geq P(X_2 \geq k), \text{ for all integers } k \geq 0. \end{aligned} \tag{10.55}$$

(Hint: For (10.54), represent X_1 using the random variables $\{B_i\}_{i=1}^{n_1}$ and represent X_2 using the first n_2 of these very same random variables. For (10.55), let $\{U_i\}_{i=1}^n$ be independent and identically distributed random variables, distributed according to the uniform distribution on $[0, 1]$; that is, $P(a \leq U_i \leq b) = b - a$, for $0 \leq a < b \leq 1$. Define random variables $\{B_i^{(1)}\}_{i=1}^n$ and $\{B_i^{(2)}\}_{i=1}^n$ by the formulas

$$B_i^{(1)} = \begin{cases} 1, & \text{if } U_i \leq p_1; \\ 0, & \text{if } U_i > p_1, \end{cases} \qquad B_i^{(2)} = \begin{cases} 1, & \text{if } U_i \leq p_2; \\ 0, & \text{if } U_i > p_2. \end{cases}$$

Now represent X_1 and X_2 through $\{B_i^{(1)}\}_{i=1}^n$ and $\{B_i^{(2)}\}_{i=1}^n$, respectively. This method is called *coupling*.)

(b) Prove (10.54) and (10.55) directly from the fact that if $X \sim \text{Bin}(n, p)$, then for $0 \leq k \leq n$, one has $P(X \geq k) = \sum_{j=k}^n \binom{n}{j} p^j (1-p)^{n-j}$.

Exercise 10.7. If $\{X_t\}_{t=0}^{\infty}$ is a Galton–Watson branching process of the type described at the beginning of Sect. 10.5, show that $EX_{t+1} = \mu EX_t$, where μ is the mean number of offspring of a particle. (Hint: Use induction and conditional expectation.)

Exercise 10.8. Prove (10.37) by the method used to prove (10.36).

Exercise 10.9. Prove that if one plays a game with three possible outcomes on each step—win, loss, or draw—with respective nonzero probabilities p', q', and r', and the outcomes of all the steps are independent of one another, and one continues to play step after step until either a win or a loss occurs, then the probability of a win is $\frac{p'}{p'+q'}$ and the probability of a loss is $\frac{q'}{p'+q'}$.

Exercise 10.10. In the proof of Theorem 10.2, after the algorithm for finding the connected component of a vertex was implemented a maximum of M_ϵ times, and a component with size around $(1-\alpha)n$ was found with probability close to 1, the final paragraph of the proof of the theorem gave a proof that with probability approaching 1 as $n \to \infty$, all other components are of size $o(n)$ as $n \to \infty$. To prove the stronger result, as in the statement of Theorem 10.2, that with probability approaching 1 as $n \to \infty$ all other components are of size $O(\log n)$, consider starting the algorithm all over again after the component of size around $(1-\alpha)n$ has been discovered. The number of edges left is around $n' = \alpha n$ and the edge probability is still $\frac{c}{n}$, which we can write as $\frac{C}{n'}$ with $C \approx c\alpha$. If $C < 1$, then the method of proof of Theorem 10.1 shows that with probability approaching 1 as $n \to \infty$ all components are of size $O(\log n') = O(\log n)$ as $n \to \infty$. To show that $C < 1$, it suffices to show that $c\alpha < 1$. To prove this, use the following facts: (1) xe^{-x} increases in $[0,1)$ and decreases in $(1,\infty)$, so for $c > 1$, there exists a unique $d \in (0,1)$ such that $de^{-d} = ce^{-c}$; (2) $\alpha = e^{c(\alpha-1)}$.

Chapter Notes

The context in which Theorems 10.1 and 10.2 were originally proven by Erdős and Rényi in 1960 [18] is a little different from the context presented here. Let $N := \binom{n}{2}$. Define $G(n,M)$, $0 \le M \le N$, to be the random graph with n vertices and exactly M edges, where the M edges are selected uniformly at random from the N possible edges. One can consider an *evolving* random graph $\{G(n,t)\}_{t=0}^{N}$. By definition, $G(n,0)$ is the graph on n vertices with no edges. Then sequentially, given $G(n,t)$, for $0 \le t \le N-1$, one obtains the graph $G(n,t+1)$ by choosing at random from the complete graph K_n one of the edges that is not in $G(n,t)$ and adjoining it to $G(n,t)$. Erdős and Rényi looked at evolving graphs of the form $G(n,t_n)$, with $t_n = [\frac{cn}{2}]$. They showed that if $c < 1$, then with probability approaching 1 as $n \to \infty$, the largest component of $G(n,t_n)$ is of size $O(\log n)$, while if $c > 1$, then with probability approaching 1 as $n \to \infty$ there is one component of size approximately $\beta(c) \cdot n$, and all other components are of size $O(\log n)$. To see how this connects up to the version given in this chapter, note that the expected number of edges in the graph $G_n(\frac{c}{n})$ is $\frac{c}{n}\binom{n}{2} = \frac{c(n-1)}{2}$. A detailed study of the borderline case, when $t_n \sim \frac{n}{2}$ as $n \to \infty$, was undertaken by Bollobás [8]. Our proofs of Theorems 10.1 and 10.2 are along the lines of the method sketched briefly in the book of Alon and Spencer [2]. We are not aware in the literature of a complete proof of Theorems 10.1 and 10.2 with all the details.

The large deviations bound in Proposition 10.2 is actually tight. That is, in part (i), where $\rho_0 > \rho$, for any $\epsilon > 0$, one has for sufficiently large n, $P(S_n \ge \rho_0 n) \ge e^{-(\kappa(\rho_0,\rho)+\epsilon)n}$. Thus, in particular, $\lim_{n\to\infty} \frac{1}{n} \log P(S_n \ge \rho_0 n) = -\kappa(\rho_0, \rho)$. Similarly, in part (ii), where $\rho_0 < \rho$, $\lim_{n\to\infty} \frac{1}{n} \log P(S_n \le \rho_0 n) = -\kappa(\rho_0, \rho)$. Consider two measures, μ and μ_0, defined on a finite or countably infinite set A. Then $H(\mu_0; \mu) := \sum_{x\in A} \mu_0(x) \log \frac{\mu_0(x)}{\mu(x)}$ is called the *relative entropy* of μ_0 with respect to μ. It plays a fundamental role in the theory of large deviations. In the case that A is a two-point set, say $A = \{0, 1\}$, and $\mu(\{1\}) = 1 - \mu(\{0\}) = \rho$ and $\mu_0(\{1\}) = 1 - \mu_0(\{0\}) = \rho_0$, one has $H(\mu_0; \mu) = \kappa(\rho_0, \rho)$. For more on large deviations, see the book by Dembo and Zeitouni [13].

For some basic results on the Galton–Watson branching process, using probabilistic methods, see the advanced probability textbook of Durrett [16]. Two standard texts on branching processes are the books of Harris [24] and of Athreya and Ney [7].

Appendix A
A Quick Primer on Discrete Probability

In this appendix, we develop some basic ideas in discrete probability theory. We note from the outset that some of the definitions given here are no longer correct in the setting of continuous probability theory.

Let Ω be a finite or countably infinite set, and let 2^Ω denote the set of subsets of Ω. An element $A \in 2^\Omega$ is simply a subset of Ω, but in the language of probability it is called an *event*. A probability measure on Ω is a function $P : 2^\Omega \to [0,1]$ satisfying $P(\emptyset) = 0$, $P(\Omega) = 1$ and which is σ*-additive*; that is, for any $1 \le N \le \infty$, one has $P(\cup_{n=1}^N A_n) = \sum_{n=1}^N P(A_n)$, whenever the events $\{A_n\}_{n=1}^N$ are disjoint. From this σ-additivity, it follows that P is uniquely determined by $\{P(\{x\})\}_{x\in\Omega}$. Using the σ-additivity on disjoint events, it is not hard to prove that P is σ-sub-additive on arbitrary events; that is, $P(\cup_{n=1}^N A_n) \le \sum_{n=1}^N P(A_n)$, for arbitrary events $\{A_n\}_{n=1}^N$. See Exercise A.1. The pair (Ω, P) is called a *probability space*.

If C and D are events and $P(C) > 0$, then the *conditional probability* of D given C is denoted by $P(D|C)$ and is defined by

$$P(D|C) = \frac{P(C \cap D)}{P(C)}.$$

Note that $P(\cdot|C)$ is itself a probability measure on Ω. Two events C and D are called *independent* if $P(C \cap D) = P(C)P(D)$. Clearly then, C and D are independent if either $P(C) = 0$ or $P(D) = 0$. If $P(C), P(D) > 0$, it is easy to check that independence is equivalent to either of the following two equalities: $P(D|C) = P(D)$ or $P(C|D) = P(C)$. Consider a collection $\{C_n\}_{n=1}^N$ of events, with $1 \le N \le \infty$. This collection of events is said to be *independent* if for any finite subset $\{C_{n_j}\}_{j=1}^m$ of the events, one has $P(\cap_{j=1}^m C_{n_j}) = \prod_{j=1}^m P(C_{n_j})$.

Let (Ω, P) be a probability space. A function $X : \Omega \to \mathbb{R}$ is called a (discrete, real-valued) *random variable*. For $B \subset \mathbb{R}$, we write $\{X \in B\}$ to denote the event $X^{-1}(B) = \{\omega \in \Omega : X(\omega) \in B\}$, the inverse image of B. When considering the probability of the event $\{X \in B\}$ or the event $\{X = x\}$, we write $P(X \in B)$ or $P(X = x)$, instead of $P(\{X \in B\})$ or $P(\{X = x\})$. The *distribution* of the random

R.G. Pinsky, *Problems from the Discrete to the Continuous*, Universitext,
DOI 10.1007/978-3-319-07965-3, © Springer International Publishing Switzerland 2014

variable X is the probability measure μ_X on $\mathbb{R}$ defined by $\mu_X(B) = P(X \in B)$, for $B \subset \mathbb{R}$. The function $p_X(x) := P(X = x)$ is called the *probability function* or the *discrete density function* for X.

The *expected value* or *expectation* EX of a random variable X is defined by

$$EX = \sum_{x\in\mathbb{R}} x\, P(X = x) = \sum_{x\in\mathbb{R}} x\, p_X(x), \text{ if } \sum_{x\in\mathbb{R}} |x|\, P(X = x) < \infty.$$

Note that the set of $x \in \mathbb{R}$ for which $P(X = x) > 0$ is either finite or countably infinite; thus, these summations are well defined. We frequently denote EX by μ. If $P(X \geq 0) = 1$ and the condition above in the definition of EX does not hold, then we write $EX = \infty$. In the sequel, when we say that the expectation of X "exists," we mean that $\sum_{x\in\mathbb{R}} |x|\, P(X = x) < \infty$.

Given a function $\psi : \mathbb{R} \to \mathbb{R}$ and a random variable X, we can define a new random variable $Y = \psi(X)$. One can calculate EY according to the definition of expectation above or in the following equivalent way:

$$EY = \sum_{x\in\mathbb{R}} \psi(x)P(X = x), \text{ if } \sum_{x\in\mathbb{R}} |\psi(x)|P(X = x) < \infty.$$

For $n \in \mathbb{N}$, the nth *moment* of X is defined by

$$EX^n = \sum_{x\in\mathbb{R}} x^n P(X = x), \text{ if } \sum_{x\in\mathbb{R}} |x|^n P(X = x) < \infty.$$

If $\mu = EX$ exists, then one defines the *variance* of X, denoted by σ^2 or $\sigma^2(X)$ or $\mathrm{Var}(X)$, by

$$\sigma^2 = E(X - \mu)^2 = \sum_{x\in\mathbb{R}} (x - \mu)^2 P(X = x).$$

Of course, it is possible to have $\sigma^2 = \infty$. It is easy to check that

$$\sigma^2(X) = EX^2 - \mu^2. \tag{A.1}$$

Chebyshev's inequality is a fundamental inequality involving the expected value and the variance.

Proposition A.1 (Chebyshev's Inequality). *Let X be a random variable with expectation μ and finite variance σ^2. Then for all $\lambda > 0$,*

$$P(|X - \mu| \geq \lambda) \leq \frac{\sigma^2}{\lambda^2}.$$

Proof.

$$P(|X-\mu|\geq\lambda)=\sum_{x\in\mathbb{R}:|x-\mu|\geq\lambda}P(X=x)\leq\sum_{x\in\mathbb{R}:|x-\mu|\geq\lambda}\frac{(x-\mu)^2}{\lambda^2}P(X=x)\leq$$

$$\sum_{x\in\mathbb{R}}\frac{(x-\mu)^2}{\lambda^2}P(X=x)=\frac{\sigma^2}{\lambda^2}.$$

□

Let $\{X_j\}_{j=1}^n$ be a finite collection of random variables on a probability space (Ω,P). We call $X=(X_1,\ldots,X_n)$ a random vector. The *joint probability function* of these random variables, or equivalently, the probability function of the random vector, is given by

$$p_X(x)=p_X(x_1,\ldots,x_n):=P(X_1=x_1,\ldots,X_n=x_n)=P(X=x),$$
$$x_i\in\mathbb{R},\ i=1,\ldots,n,\ \text{where } x=(x_1,\ldots,x_n).$$

It follows that $\sum_{j\in[n]-\{i\}}\sum_{x_j\in\mathbb{R}}p_X(x)=P(X_i=x_i)$. For any function $H: R^n\to R$, we define

$$EH(X)=\sum_{x\in\mathbb{R}^n}H(x)p_X(x),\ \text{if}\ \sum_{x\in\mathbb{R}^n}|H(x)|p_X(x)<\infty.$$

In particular then, if EX_j exists, it can be written as $EX_j=\sum_{x\in\mathbb{R}^n}x_jp_X(x)$. Similarly, if EX_k exists, for all k, then we have

$$E\sum_{k=1}^n c_kX_k=\sum_{x\in\mathbb{R}^n}(\sum_{k=1}^n c_kx_k)p_X(x)=\sum_{k=1}^n c_k(\sum_{x\in\mathbb{R}^n}x_kp_X(x)).$$

It follows from this that the expectation is linear; that is, if EX_k exists for $k=1,\ldots,n$, then

$$E\sum_{k=1}^n c_kX_k=\sum_{k=1}^n c_kEX_k,$$

for any real numbers $\{c_k\}_{k=1}^n$.

Let $\{X_j\}_{j=1}^N$ be a collection of random variables on a probability space (Ω,P), where $1\leq N\leq\infty$. The random variables are called *independent* if for every finite $n\leq N$, one has

$$P(X_1 = x_1, X_2 = x_2, \ldots, X_n = x_n) = \prod_{j=1}^{n} P(X_j = x_j),$$

for all $x_j \in \mathbb{R}$, $j = 1, 2, \ldots, n$.

Let $\{f_i\}_{i=1}^n$ be real-valued functions with f_i defined at least on the set $\{x \in \mathbb{R} : P(X_i = x) > 0\}$. Assume that $E|f_i(X_i)| < \infty$, for $i = 1, \ldots, n$. From the definition of independence it is easy to show that if $\{X_j\}_{j=1}^n$ are independent, then

$$E \prod_{i=1}^{n} f_i(X_i) = \prod_{i=1}^{n} E f_i(X_i). \tag{A.2}$$

The variance is of course not linear. However the variance of a sum of independent random variables is equal to the sum of the variances of the random variables:

If $\{X_i\}_{i=1}^n$ are independent random variables, then

$$\sigma^2(\sum_{i=1}^{n} X_i) = \sum_{i=1}^{n} \sigma^2(X_i). \tag{A.3}$$

It suffices to prove (A.3) for $n = 2$ and then use induction. Let $\mu_i = EX_i, i = 1, 2$. We have

$$\sigma^2(X_1 + X_2) = E\big(X_1 + X_2 - E(X_1 + X_2)\big)^2 = E\big((X_1 - \mu_1) + (X_2 - \mu_2)\big)^2 =$$
$$E(X_1 - \mu_1)^2 + E(X_2 - \mu_2)^2 + 2E(X_1 - \mu_1)(X_2 - \mu_2) = \sigma^2(X_1) + \sigma^2(X_2),$$

where the last equality follows because (A.2) shows that $E(X_1 - \mu_1)(X_2 - \mu_2) = E(X_1 - \mu_1)E(X_2 - \mu_2) = 0$.

Chebyshev's inequality and (A.3) allow for an exceedingly short proof of an important result—*the weak law of large numbers* for sums of independent, identically distributed (IID) random variables.

Theorem A.1. *Let $\{X_n\}_{n=1}^\infty$ be a sequence of independent, identically distributed random variables and assume that their common variance σ^2 is finite. Denote their common expectation by μ. Let $S_n = \sum_{j=1}^n X_j$. Then for any $\epsilon > 0$,*

$$\lim_{n \to \infty} P(|\frac{S_n}{n} - \mu| \geq \epsilon) = 0.$$

Proof. We have $ES_n = n\mu$, and since the random variables are independent and identically distributed, it follows from (A.3) that $\sigma^2(S_n) = n\sigma^2$. Now applying Chebyshev's inequality to S_n with $\lambda = n\epsilon$ gives

$$P(|S_n - n\mu| \geq n\epsilon) \leq \frac{n\sigma^2}{(n\epsilon)^2},$$

which proves the theorem. □

Remark. The weak law of large numbers is a first moment result. It holds even without the finite variance assumption, but the proof is much more involved.

The above weak law of large numbers is actually a particular case of the following weak law of large numbers.

Proposition A.2. *Let $\{Y_n\}_{n=1}^{\infty}$ be random variables. Assume that*

$$\sigma^2(Y_n) = o\big((EY_n)^2\big), \text{ as } n \to \infty.$$

Then for any $\epsilon > 0$,

$$\lim_{n\to\infty} P(|\frac{Y_n}{EY_n} - 1| \geq \epsilon) = 0.$$

Proof. By Chebyshev's inequality, we have

$$P(|Y_n - EY_n| \geq \epsilon|EY_n|) \leq \frac{\sigma^2(Y_n)}{(\epsilon EY_n)^2}.$$

□

If X and Y are random variables on a probability space (Ω, P), and if $P(X{=}x){>}0$, then the *conditional probability function of Y given $X = x$* is defined by

$$p_{Y|X}(y|x) := P(Y = y|X = x) = \frac{P(X = x, Y = y)}{P(X = x)}.$$

The *conditional expectation of Y given $X = x$* is defined by

$$E(Y|X = x) = \sum_{y\in\mathbb{R}} y\,P(Y = y|X = x) = \sum_{y\in\mathbb{R}} y\; p_{Y|X}(y|x),$$

$$\text{if } \sum_{y\in\mathbb{R}} |y| P(Y = y|X = x) < \infty.$$

It is easy to verify that

$$EY = \sum_{x\in\mathbb{R}} E(Y|X = x)P(X = x),$$

where $E(Y|X = x)P(X = x) := 0$, if $P(X = x) = 0$.

A random variable X that takes on only two values—0 and 1, with $P(X = 1) = p$ and $P(X = 0) = 1 - p$, for some $p \in [0, 1]$—is called a *Bernoulli random variable*. One writes $X \sim \text{Ber}(p)$. It is trivial to check that $EX = p$ and $\sigma^2(X) = p(1 - p)$.

Let $n \in \mathbb{N}$ and let $p \in [0, 1]$. A random variable X satisfying

$$P(X = j) = \binom{n}{j} p^j (1 - p)^{n-j}, \ j = 0, 1, \ldots, n,$$

is called a *binomial random variable*, and one writes $X \sim \text{Bin}(n, p)$. The random variable X can be thought of as the number of "successes" in n independent trials, where on each trial there are two possible outcomes—"success" and "failure"—and the probability of "success" is p on each trial. Letting $\{Z_i\}_{i=1}^n$ be independent, identically distributed random variables distributed according to Ber(p), it follows that X can be realized as $X = \sum_{i=1}^n Z_i$. From the formula for the expected value and variance of a Bernoulli random variable, and from the linearity of the expectation and (A.3), the above representation immediately yields $EX = np$ and $\sigma^2(X) = np(1 - p)$.

A random variable X satisfying

$$P(X = n) = e^{-\lambda} \frac{\lambda^n}{n!}, \ n = 0, 1, \ldots,$$

where $\lambda > 0$, is called a *Poisson random variable*, and one writes $X \sim \text{Pois}(\lambda)$. One can check easily that $EX = \lambda$ and $\sigma^2(X) = \lambda$.

Proposition A.3 (Poisson Approximation to the Binomial Distribution). *For $n \in \mathbb{N}$ and $p \in [0, 1]$, let $X_{n,p} \sim \text{Bin}(n, p)$. For $\lambda > 0$, let $X_\lambda \sim \text{Pois}(\lambda)$. Then*

$$\lim_{n \to \infty, p \to 0, np \to \lambda} P(X_{n,p} = j) = P(X_\lambda = j), \ j = 0, 1, \ldots. \tag{A.4}$$

Proof. By assumption, we have $p = \frac{\lambda_n}{n}$, where $\lim_{n\to\infty} \lambda_n = \lambda$. We have

$$P(X_{n,p}=j)=\binom{n}{j} p^j (1-p)^{n-j} = \frac{n(n-1)\cdots(n-j+1)}{j!} (\frac{\lambda_n}{n})^j (1 - \frac{\lambda_n}{n})^{n-j} =$$

$$\frac{1}{j!} \lambda_n^j \frac{n(n-1)\cdots(n-j+1)}{n^j} (1 - \frac{\lambda_n}{n})^{n-j};$$

thus,

$$\lim_{n \to \infty, p \to 0, np \to \lambda} P(X_{n,p} = j) = e^{-\lambda} \frac{\lambda^j}{j!} = P(X_\lambda = j).$$

□

Equation (A.4) is an example of *weak convergence* of random variables or distributions. In general, if $\{X_n\}_{n=1}^{\infty}$ are random variables with distributions $\{\mu_{X_n}\}_{n=1}^{\infty}$, and X is a random variable with distribution μ_X, then we say that X_n *converges weakly to* X, or μ_{X_n} *converges weakly to* μ_X, if $\lim_{n\to\infty} P(X_n \le x) = P(X \le x)$, for all $x \in \mathbb{R}$ for which $P(X = x) = 0$, or equivalently, if $\lim_{n\to\infty} \mu_{X_n}((-\infty, x]) = \mu_X((-\infty, x])$, for all $x \in \mathbb{R}$ for which $\mu_X(\{x\}) = 0$. Thus, for example, if $P(X_n = \frac{1}{n}) = P(X_n = 1 + \frac{1}{n}) = \frac{1}{2}$, for $n = 1, 2, \cdots$, and $P(X = 0) = P(X = 1) = \frac{1}{2}$, then X_n converges weakly to X since $\lim_{n\to\infty} P(X_n \le x) = P(X \le x)$, for all $x \in \mathbb{R} - \{0, 1\}$. See also Exercise A.4.

Exercise A.1. Use the σ-additivity property of probability measures on disjoint sets to prove σ-sub-additivity on arbitrary sets: that is, $P(\cup_{n=1}^{N} A_n) \le \sum_{n=1}^{N} P(A_n)$, for arbitrary events $\{A_n\}_{n=1}^{N}$, where $1 \le N \le \infty$. (Hint: Rewrite $\cup_{n=1}^{N} A_n$ as a disjoint union $\cup_{n=1}^{N} B_n$, by letting $B_1 = A_1$, $B_2 = A_2 - A_1$, $B_3 = A_3 - A_2 - A_1$, etc.)

Exercise A.2. Prove that $P(A_1 \cup A_2) = P(A_1) + P(A_2) - P(A_1 \cap A_2)$, for arbitrary events A_1, A_2. Then prove more generally that for any finite n and arbitrary events $\{A_k\}_{k=1}^{n}$, one has

$$P(\cup_{k=1}^{n} A_k) = \sum_{1\le i\le n} P(A_i) - \sum_{1\le i<j\le n} P(A_i \cap A_j) +$$

$$\sum_{1\le i<j<k\le n} P(A_i \cap A_j \cap A_k) - \cdots + (-1)^{n-1} P(A_1 \cap A_2 \cdots \cap A_n).$$

This result is known as the *principle of inclusion–exclusion*.

Exercise A.3. Let (Ω, P) be a probability space and let $R \ge 2$ be an integer. For $A \subset \Omega$, recall that the complement A^c of A is defined by $A^c = \Omega - A$. Prove that if the events $\{A_k\}_{k=1}^{R}$ are independent, then the complementary events $\{A_k^c\}_{k=1}^{R}$ are also independent. (Hint: By the definition of independence, we have

$$P(\cap_{j=1}^{\ell} B_j) = \prod_{j=1}^{\ell} P(B_j), \text{ for any } \ell \le R \text{ and any sub-collection } \{B_j\}_{j=1}^{\ell} \text{ of } \{A_k\}_{k=1}^{R}. \tag{A.5}$$

Using this, we need to prove that $P(\cap_{j=1}^{\ell} B_j^c) = \prod_{j=1}^{\ell} P(B_j^c)$, for any sub-collection $\{B_j^c\}_{j=1}^{\ell}$ of $\{A_k^c\}_{k=1}^{R}$. Let $p_j = P(B_j)$ and $p = P(\cap_{j=1}^{\ell} B_j^c)$. Then we need to prove that $p = \prod_{j=1}^{\ell}(1 - p_j)$. Write

$$\prod_{j=1}^{\ell}(1-p_j) = 1 - \sum_{1\le i\le \ell} p_i + \sum_{1\le i<j\le \ell} p_i p_j - \cdots,$$

and use (A.5) along with the principle of inclusion–exclusion, which appears in Exercise A.2.

Exercise A.4. Using (A.4), show that

$$\lim_{n\to\infty, p\to 0, np\to\lambda} P(X_{n,p} \le x) = P(X_\lambda \le x), \text{ for all } x \in R.$$

Appendix B
Power Series and Generating Functions

We review without proof some basic results concerning power series. For more details, the reader should consult an advanced calculus or undergraduate analysis text. We also illustrate the utility of generating functions by analyzing the one that arises from the Fibonacci sequence.

Let $\{a_n\}_{n=0}^{\infty}$ be a sequence of real numbers. Define formally the *generating function* $F(t)$ of $\{a_n\}_{n=0}^{\infty}$ by

$$F(t) = \sum_{n=0}^{\infty} a_n t^n, \tag{B.1}$$

where $t \in R$. We say "formally" because we have made the definition before determining for which values of t the power series on the right hand side above converges. The power series converges trivially for $t = 0$, and it is possible that it converges only for $t = 0$, for example, if $a_n = n!$.

The power series $\sum_{n=0}^{\infty} a_n t^n$ converges *absolutely* if $\sum_{n=0}^{\infty} |a_n t^n| < \infty$. The power series is *uniformly, absolutely convergent* for $|t| \leq \rho$ if

$$\lim_{N\to\infty} \sup_{|t|\leq\rho} \sum_{n=N}^{\infty} |a_n t^n| = 0;$$

that is, if the tail of the series $\sum_{n=0}^{\infty} |a_n t^n|$ converges to 0 uniformly over $|t| \leq \rho$.

We state four fundamental results concerning the convergence of power series:

1. *If the power series converges for some number $t_0 \neq 0$, then necessarily the power series converges absolutely and uniformly for $|t| \leq \rho$, for all $\rho < t_0$.*
2. *There exists an extended real number $r_0 \in [0, \infty]$ such that the power series $\sum_{n=0}^{\infty} a_n t^n$ converges absolutely if $t \in [0, r_0)$ and diverges if $t > r_0$.*
   The number r_0 in (2) is called the *radius of convergence* of the power series.

R.G. Pinsky, *Problems from the Discrete to the Continuous*, Universitext,
DOI 10.1007/978-3-319-07965-3, © Springer International Publishing Switzerland 2014

3. *The radius of convergence is given by the formula*

$$r_0 = \frac{1}{\limsup_{n\to\infty} \sqrt[n]{a_n}}.$$

4. *If the power series is uniformly, absolutely convergent for* $|t| \le \rho$, *then the function* $F(t)$ *in* (B.1) *is infinitely differentiable for* $|t| < \rho$, *and its derivatives are obtained via term by term differentiation in the power series; in particular,* $F'(t) = \sum_{n=0}^{\infty} n a_n t^{n-1}$.

The generating function often provides an efficient method for obtaining information about the sequence $\{a_n\}_{n=0}^{\infty}$. Typically, this will occur when the generating function can be written in a nice closed form and analyzed. This analysis then allows one to obtain information about the coefficients in the generating function's power series expansion, and these coefficients are of course $\{a_n\}_{n=0}^{\infty}$. We illustrate this in the case of the famous *Fibonacci sequence.*

Recall that the sequence of Fibonacci numbers is defined recursively by $f_0 = 0$, $f_1 = 1$ and

$$f_n = f_{n-1} + f_{n-2}, \text{ for } n \ge 2. \tag{B.2}$$

The first few Fibonacci numbers are 0,1,1,2,3,5,8,13, 21,34,55,89,144.

We will obtain a closed form for the generating function

$$F(t) = \sum_{n=0}^{\infty} f_n t^n \tag{B.3}$$

of the Fibonacci numbers. Multiply both sides of (B.2) by t^n and then sum both sides over n, with n running from 2 to ∞. This gives us

$$\sum_{n=2}^{\infty} f_n t^n = \sum_{n=2}^{\infty} f_{n-1} t^n + \sum_{n=2}^{\infty} f_{n-2} t^n.$$

Since $f_0 = 0$ and $f_1 = 1$, the left hand side above is equal to $F(t) - t$. Factoring out t from the first term and t^2 from the second term on the right hand side above, and using the fact that $f_0 = 0$, one sees that the right hand side above is equal to $tF(t) + t^2F(t)$. Thus, we obtain the equation

$$F(t) - t = tF(t) + t^2F(t),$$

which gives a closed form expression for F; namely, $F(t) = \frac{t}{1-t-t^2}$. Up until now we have ignored the question of convergence. However, the above formula gives us the answer. The roots of the polynomial $t^2 + t - 1$ are $r^+ := \frac{-1+\sqrt{5}}{2}$ and $r^- := \frac{-1-\sqrt{5}}{2}$. Since $|r^+| < |r^-|$, we conclude that the generating function $F(t)$ has radius of convergence $|r^+| = \frac{\sqrt{5}-1}{2}$. Thus, the generating function of the Fibonacci series is given by

$$F(t) = \frac{t}{1-t-t^2}, \quad |t| < \frac{\sqrt{5}-1}{2}. \tag{B.4}$$

We now use the method of partial fractions to represent the function $\frac{t}{1-t-t^2}$ in an explicit power series. Using the fact that $r^+r^- = -1$, we write

$$t^2 + t - 1 = (t - r^+)(t - r^-) = -(tr^- + 1)(tr^+ + 1);$$

thus,

$$\frac{t}{1-t-t^2} = \frac{t}{(tr^- + 1)(tr^+ + 1)}. \tag{B.5}$$

For unknown A and B, we write

$$\frac{t}{(tr^- + 1)(tr^+ + 1)} = \frac{A}{tr^- + 1} + \frac{B}{tr^+ + 1} = \frac{t(Ar^+ + Br^-) + (A + B)}{(tr^- + 1)(tr^+ + 1)}. \tag{B.6}$$

Comparing the left-most and right-most terms in (B.6) , we conclude that $A+B = 0$ and $Ar^+ + Br^- = 1$. Solving for A and B, we obtain $A = \frac{1}{r^+ - r^-} = \frac{1}{\sqrt{5}}$ and $B = \frac{1}{r^- - r^+} = -\frac{1}{\sqrt{5}}$. Thus, from (B.5) and the first equality in (B.6), we arrive at the partial fraction representation

$$\frac{t}{1-t-t^2} = \frac{1}{\sqrt{5}}\left(\frac{1}{1+tr^-} - \frac{1}{1+tr^+}\right). \tag{B.7}$$

Since $|r^-| > |r^+|$, both $\frac{1}{1+tr^-}$ and $\frac{1}{1+tr^+}$ can be written as geometric series if $|t| < \frac{1}{|r^-|} = \frac{2}{1+\sqrt{5}} = \frac{\sqrt{5}-1}{2}$. We have

$$\begin{aligned} \frac{1}{1+tr^-} &= \sum_{n=0}^{\infty}(-1)^n(r^-)^n t^n = \sum_{n=0}^{\infty}\left(\frac{1+\sqrt{5}}{2}\right)^n t^n; \\ \frac{1}{1+tr^+} &= \sum_{n=0}^{\infty}(-1)^n(r^+)^n t^n = \sum_{n=0}^{\infty}\left(\frac{1-\sqrt{5}}{2}\right)^n t^n. \end{aligned} \tag{B.8}$$

Thus, from (B.4), (B.7), and (B.8), we obtain

$$F(t) = \sum_{n=0}^{\infty} \frac{1}{\sqrt{5}}\left(\left(\frac{1+\sqrt{5}}{2}\right)^n - \left(\frac{1-\sqrt{5}}{2}\right)^n\right)t^n. \tag{B.9}$$

Comparing (B.3) with (B.9), we conclude that the nth Fibonacci number f_n is given explicitly by

$$f_n = \frac{1}{\sqrt{5}}\Big((\frac{1+\sqrt{5}}{2})^n - (\frac{1-\sqrt{5}}{2})^n\Big). \tag{B.10}$$

From the explicit formula in (B.10), the asymptotic behavior of f_n is clear:

$$f_n \sim \frac{1}{\sqrt{5}}(\frac{1+\sqrt{5}}{2})^n \text{ as } n \to \infty.$$

Appendix C
A Proof of Stirling's Formula

Stirling's formula states that

$$n! \sim n^n e^{-n} \sqrt{2\pi n}, \text{ as } n \to \infty. \tag{C.1}$$

In order to obtain an asymptotic formula for the discrete quantity $n!$, it is extremely useful to be able to embed this quantity in a function of a continuous variable. Integrating by parts and then applying induction shows that $n! = \Gamma(n+1)$, $n \in \mathbb{N}$, where the *gamma function* $\Gamma(t)$ is defined by

$$\Gamma(t) = \int_0^\infty x^{t-1} e^{-x}\, dx, \quad t > 0.$$

Thus, one proves Stirling's formula in the following form.

Theorem C.1 (Stirling's Formula).

$$\Gamma(t+1) \sim t^t e^{-t} \sqrt{2\pi t}, \textit{ as } t \to \infty. \tag{C.2}$$

Proof. In the literature one can find literally dozens of proofs of Stirling's formula. We present here an elementary proof that uses *Laplace's asymptotic method* [14]. We begin by giving the intuition for the method. We write

$$\Gamma(t+1) = \int_0^\infty e^{\psi_t(x)}\, dx, \tag{C.3}$$

where

$$\psi_t(x) = t \log x - x.$$

R.G. Pinsky, *Problems from the Discrete to the Continuous*, Universitext,
DOI 10.1007/978-3-319-07965-3, © Springer International Publishing Switzerland 2014

Now ψ_t takes on its maximum at $x = t$, and the Taylor expansion of ψ_t about $x = t$ starts out as

$$t \log t - t - \frac{(x-t)^2}{2t} =: \hat{\psi}_t(x).$$

Replacing ψ_t by $\hat{\psi}_t$, we calculate that

$$\int_0^\infty e^{\hat{\psi}_t(x)}\,dx = \int_0^\infty e^{t \log t - t - \frac{(x-t)^2}{2t}}\,dx = t^t e^{-t} \int_0^\infty e^{-\frac{(x-t)^2}{2t}}\,dx.$$

Making the substitution $z = \frac{x-t}{\sqrt{t}}$ gives

$$\int_0^\infty e^{-\frac{(x-t)^2}{2t}}\,dx = \sqrt{t} \int_{-\sqrt{t}}^\infty e^{-\frac{1}{2}z^2}\,dz.$$

Since $\int_{-\infty}^\infty e^{-\frac{1}{2}z^2}\,dz = \sqrt{2\pi}$, we conclude that

$$\int_0^\infty e^{\hat{\psi}_t(x)}\,dx \sim t^t e^{-t} \sqrt{2\pi t}, \text{ as } t \to \infty.$$

We now turn to the rigorous proof. We can write ψ_t exactly as

$$\psi_t(t+y) = t \log t - t - tg(\frac{y}{t}),$$

where

$$g(v) = v - \log(1+v).$$

Substituting this in (C.3) and making the change of variables $x = y + t$, we obtain

$$\Gamma(t+1) = t^t e^{-t} \int_{-t}^\infty e^{-tg(\frac{y}{t})}\,dy.$$

Making the change of variables $y = \sqrt{t}z$, we have

$$\Gamma(t+1) = t^t e^{-t} \sqrt{2\pi t}\, \bar{\Gamma}(t), \tag{C.4}$$

where

$$\bar{\Gamma}(t) = \frac{1}{\sqrt{2\pi}} \int_{-\sqrt{t}}^\infty e^{-tg(\frac{z}{\sqrt{t}})}\,dz.$$

We will show that

$$\lim_{t\to\infty} \bar{\Gamma}(t) = 1. \tag{C.5}$$

Now (C.2) follows from (C.4) and (C.5).

Fix $L > 0$ and write

$$\bar{\Gamma}(t) = \bar{\Gamma}_L(t) + \frac{1}{\sqrt{2\pi}} T_L^+(t) + \frac{1}{\sqrt{2\pi}} T_L^-(t), \tag{C.6}$$

where

$$\bar{\Gamma}_L(t) = \frac{1}{\sqrt{2\pi}} \int_{-L}^{L} e^{-tg(\frac{z}{\sqrt{t}})}\, dz$$

and

$$T_L^+(t) = \int_L^\infty e^{-tg(\frac{z}{\sqrt{t}})}\, dz, \quad T_L^-(t) = \int_{-\sqrt{t}}^{-L} e^{-tg(\frac{z}{\sqrt{t}})}\, dz.$$

From Taylor's remainder formula it follows that for any $\epsilon > 0$ and sufficiently small v, one has

$$\frac{1}{2}(1-\epsilon)v^2 \le g(v) \le \frac{1}{2}(1+\epsilon)v^2.$$

Thus, $\lim_{t\to\infty} tg(\frac{z}{\sqrt{t}}) = \frac{1}{2}z^2$, uniformly over $z \in [-L, L]$; consequently,

$$\lim_{t\to\infty} \bar{\Gamma}_L(t) = \frac{1}{\sqrt{2\pi}} \int_{-L}^{L} e^{-\frac{1}{2}z^2}\, dz. \tag{C.7}$$

Since $t\big(g(\frac{z}{\sqrt{t}})\big)' = \sqrt{t}\big(1 - \frac{1}{1+\frac{z}{\sqrt{t}}}\big) = \frac{\sqrt{t}z}{\sqrt{t}+z}$ is increasing in z, we have

$$T_L^+(t) \le \frac{\sqrt{t}+L}{\sqrt{t}L} \int_L^\infty t\big(g(\frac{z}{\sqrt{t}})\big)' e^{-tg(\frac{z}{\sqrt{t}})}\, dz = \frac{\sqrt{t}+L}{\sqrt{t}L} e^{-tg(\frac{L}{\sqrt{t}})} =$$

$$\frac{\sqrt{t}+L}{\sqrt{t}L} e^{-t[\frac{L}{\sqrt{t}} - \log(1+\frac{L}{\sqrt{t}})]}.$$

By Taylor's formula, we have $\log(1+\frac{L}{\sqrt{t}}) = \frac{L}{\sqrt{t}} - \frac{L^2}{2t} + O(t^{-\frac{3}{2}})$ as $t \to \infty$; thus,

$$\limsup_{t\to\infty} T_L^+(t) \le \frac{1}{L} e^{-\frac{1}{2}L^2}. \tag{C.8}$$

A very similar argument gives

$$\limsup_{t\to\infty} T_L^-(t) \le \frac{1}{L} e^{-\frac{1}{2}L^2}. \tag{C.9}$$

Now from (C.6)–(C.9), we obtain

$$\begin{aligned}\frac{1}{\sqrt{2\pi}} \int_{-L}^{L} e^{-\frac{1}{2}z^2}\, dz &\le \liminf_{t\to\infty} \bar{\Gamma}(t) \le \limsup_{t\to\infty} \bar{\Gamma}(t) \\ &\le \frac{1}{\sqrt{2\pi}} \int_{-L}^{L} e^{-\frac{1}{2}z^2}\, dz + \frac{2}{L\sqrt{2\pi}} e^{-\frac{1}{2}L^2}.\end{aligned}$$

Since $\bar{\Gamma}(t)$ is independent of L, letting $L \to \infty$ above gives (C.5). □

Appendix D
An Elementary Proof of $\sum_{n=1}^{\infty} \frac{1}{n^2} = \frac{\pi^2}{6}$

The standard way to prove the identity in the title of this appendix is via Fourier series. We give a completely elementary proof, following [1]. Consider the double integral

$$I = \int_0^1 \int_0^1 \frac{1}{1-xy}\, dxdy. \tag{D.1}$$

(Actually, the expression on the right hand side of (D.1) is an improper integral, because the integrand blows up at $(x, y) = (1, 1)$. Thus, $\int_0^1 \int_0^1 \frac{1}{1-xy}\, dxdy :=$ $\lim_{\epsilon \to 0+} \int_0^{1-\epsilon} \int_0^{1-\epsilon} \frac{1}{1-xy}\, dxdy$. Since the integrand is nonnegative, there is no problem applying the standard rules of calculus directly to $\int_0^1 \int_0^1 \frac{1}{1-xy}\, dxdy$.) On the one hand, expanding the integrand in a geometric series and integrating term by term gives

$$I = \int_0^1 \int_0^1 \sum_{n=0}^{\infty} (xy)^n\, dxdy = \sum_{n=0}^{\infty} \int_0^1 \int_0^1 x^n y^n\, dxdy =$$

$$\sum_{n=0}^{\infty} \Big(\int_0^1 x^n\, dx\Big)\Big(\int_0^1 y^n\, dy\Big) = \sum_{n=0}^{\infty} \frac{1}{(n+1)^2} = \sum_{n=1}^{\infty} \frac{1}{n^2}. \tag{D.2}$$

(The interchanging of the order of the integration and the summation is justified by the fact that all the summands are nonnegative.)

On the other hand, consider the change of variables $u = \frac{y+x}{2}$, $v = \frac{y-x}{2}$. This transformation rotates the square $[0, 1] \times [0, 1]$ clockwise by $45°$ and shrinks its sides by the factor $\sqrt{2}$. The new domain is $\{(u, v) : 0 \le u \le \frac{1}{2}, -u \le v \le u\} \cup \{(u, v) :$ $\frac{1}{2} \le u \le 1, u \quad 1 \le v \le 1-u\}$. The Jacobian $\frac{\partial(x,y)}{\partial(u,v)}$ of the transformation is equal to 2, so the area element $dxdy$ gets replaced by $2dudv$. The function $\frac{1}{1-xy}$ becomes $\frac{1}{1-u^2+v^2}$. Since the function and the domain are symmetric with respect to the u-axis, we have

R.G. Pinsky, *Problems from the Discrete to the Continuous*, Universitext,
DOI 10.1007/978-3-319-07965-3, © Springer International Publishing Switzerland 2014

$$I = 4\int_0^{\frac{1}{2}} \Big(\int_0^u \frac{dv}{1-u^2+v^2}\Big)\,du + 4\int_{\frac{1}{2}}^1 \Big(\int_0^{1-u} \frac{dv}{1-u^2+v^2}\Big)\,du.$$

Using the integration formula $\int \frac{dx}{x^2+a^2} = \frac{1}{a}\arctan\frac{x}{a}$, we obtain

$$I = 4\int_0^{\frac{1}{2}} \frac{1}{\sqrt{1-u^2}}\arctan\Big(\frac{u}{\sqrt{1-u^2}}\Big)\,du + 4\int_{\frac{1}{2}}^1 \frac{1}{\sqrt{1-u^2}}\arctan\Big(\frac{1-u}{\sqrt{1-u^2}}\Big)\,du.$$

Now the derivative of $g(u) := \arctan\big(\frac{u}{\sqrt{1-u^2}}\big)$ is $\frac{1}{\sqrt{1-u^2}}$, and the derivative of $h(u) := \arctan\big(\frac{1-u}{\sqrt{1-u^2}}\big) = \arctan\big(\sqrt{\frac{1-u}{1+u}}\big)$ is $-\frac{1}{2}\frac{1}{\sqrt{1-u^2}}$. Thus, we conclude that

$$\begin{aligned}
I &= 4\int_0^{\frac{1}{2}} g(u)g'(u)\,du - 8\int_{\frac{1}{2}}^1 h(u)h'(u)\,du = 2g^2(u)|_0^{\frac{1}{2}} - 4h^2(u)|_{\frac{1}{2}}^1 = \\
&2\big(\arctan^2\frac{1}{\sqrt{3}} - \arctan^2 0\big) - 4\big(\arctan^2 0 - \arctan^2\frac{1}{\sqrt{3}}\big) = 6\arctan^2\frac{1}{\sqrt{3}} \\
&= 6\Big(\frac{\pi}{6}\Big)^2 = \frac{\pi^2}{6}. \qquad \text{(D.3)}
\end{aligned}$$

Comparing (D.2) and (D.3) gives

$$\sum_{n=1}^{\infty} \frac{1}{n^2} = \frac{\pi^2}{6}.$$

□

References

1. Aigner, M., Ziegler, G.: Proofs from the Book, 4th edn. Springer, Berlin (2010)
2. Alon, N., Spencer, J.: The Probabilistic Method, 3rd edn. Wiley-Interscience Series in Discrete Mathematics and Optimization. Wiley, Hoboken (2008)
3. Alon, N., Krivelevich, M., Sudakov, B.: Finding a large hidden clique in a random graph. In: Proceedings of the 9th Annual ACM-SIAM Symposium on Discrete Algorithms (San Francisco, CA, 1998), pp. 594–598. ACM, New York (1998)
4. Andrews, G.: The Theory of Partitions, reprint of the 1976 original. Cambridge University Press, Cambridge (1998)
5. Apostol, T.: Introduction to Analytic Number Theory. Undergraduate Texts in Mathematics. Springer, New York (1976)
6. Arratia, R., Barbour, A.D., Tavaré, S.: Logarithmic Combinatorial Structures: A Probabilistic Approach. EMS Monographs in Mathematics. European Mathematical Society, Zürich (2003)
7. Athreya, K., Ney, P.: Branching Processes, reprint of the 1963 original [Springer, Berlin]. Dover Publications, Inc., Mineola (2004)
8. Bollobás, B.: The evolution of random graphs. Trans. Am. Math. Soc. **286**, 257–274 (1984)
9. Bollobás, B.: Modern Graph Theory. Graduate Texts in Mathematics, vol. 184. Springer, New York (1998)
10. Bollobás, B.: Random Graphs, 2nd edn. Cambridge Studies in Advanced Mathematics, vol. 73. Cambridge University Press, Cambridge (2001)
11. Brauer, A.: On a problem of partitions. Am. J. Math. **64**, 299–312 (1942)
12. Conlon, D.: A new upper bound for diagonal Ramsey numbers. Ann. Math. **170**, 941–960 (2009)
13. Dembo, A., Zeitouni, O.: Large Deviations Techniques and Applications, 2nd edn. Springer, New York (1998)
14. Diaconis, P., Freedman, D.: An elementary proof of Stirling's formula. Am. Math. Mon. **93**, 123–125 (1986)
15. Doyle, P., Snell, J.L.: Random Walks and Electric Networks. Carus Mathematical Monographs, vol. 22. Mathematical Association of America, Washington (1984)
16. Durrett, R.: Probability: Theory and Examples, 4th edn. Cambridge Series in Statistical and Probabilistic Mathematics. Cambridge University Press, Cambridge (2010)
17. Dwass, M.: The number of increases in a random permutation. J. Combin. Theor. Ser. A **15**, 192–199 (1973)
18. Erdős, P., Rényi, A.: On the evolution of random graphs. Magyar Tud. Akad. Mat. Kutató Int. Közl **5**, 17–61 (1960)
19. Feller, W.: An Introduction to Probability Theory and Its Applications, 3rd edn, vol. I. Wiley, New York (1968)

R.G. Pinsky, *Problems from the Discrete to the Continuous*, Universitext,
DOI 10.1007/978-3-319-07965-3, © Springer International Publishing Switzerland 2014

20. Flajolet, P., Sedgewick, R.: Analytic Combinatorics. Cambridge University Press, Cambridge (2009)
21. Flory, P.J.: Intramolecular reaction between neighboring substituents of vinyl polymers. J. Am. Chem. Soc. **61**, 1518–1521 (1939)
22. Graham, R., Rothschild, B., Spencer, J.: Ramsey Theory, 2nd edn. Wiley-Interscience Series in Discrete Mathematics and Optimization. Wiley, New York (1990)
23. Hardy, G.H., Ramanujan, S.: Asymptotic formulae in combinatory analysis. Proc. London Math. Soc. **17**, 75–115 (1918)
24. Harris, T.: The Theory of Branching Processes, corrected reprint of the 1963 original [Springer, Berlin]. Dover Publications, Inc., Mineola (2002)
25. Jameson, G.J.O.: The Prime Number Theorem. London Mathematical Society Student Texts, vol. 53. Cambridge University Press, Cambridge (2003)
26. Montgomery, H., Vaughan, R.: Multiplicative Number Theory. I. Classical Theory. Cambridge Studies in Advanced Mathematics, vol. 97. Cambridge University Press, Cambridge (2007)
27. Nathanson, M.: Elementary Methods in Number Theory. Graduate Texts in Mathematics, vol. 195. Springer, New York (2000)
28. Page, E.S.: The distribution of vacancies on a line. J. Roy. Stat. Soc. Ser. B **21**, 364–374 (1959)
29. Pinsky, R.: Detecting tampering in a random hypercube. Electron. J. Probab. **18**, 1–12 (2013)
30. Pitman, J.: Combinatorial stochastic processes. Lectures from the 32nd Summer School on Probability Theory held in Saint-Flour, 7–24 July 2002. Lecture Notes in Mathematics, 1875. Springer, Berlin (2006)
31. Rényi, A.: On a one-dimensional problem concerning random space filling (Hungarian; English summary). Magyar Tud. Akad. Mat. Kutató Int. Közl. **3**, 109–127 (1958)
32. Spitzer, F.: Principles of Random Walk, 2nd edn. Graduate Texts in Mathematics, vol. 34. Springer, New York (1976)
33. Tenenbaum, G.: Introduction to Analytic and Probabilistic Number Theory. Cambridge Studies in Advanced Mathematics, vol. 46. Cambridge University Press, Cambridge (1995)
34. Wilf, H.: Generating Functionology, 3rd edn. A K Peters, Ltd., Wellesley (2006)

Index

R.G. Pinsky, *Problems from the Discrete to the Continuous*, Universitext,
DOI 10.1007/978-3-319-07965-3, © Springer International Publishing Switzerland 2014